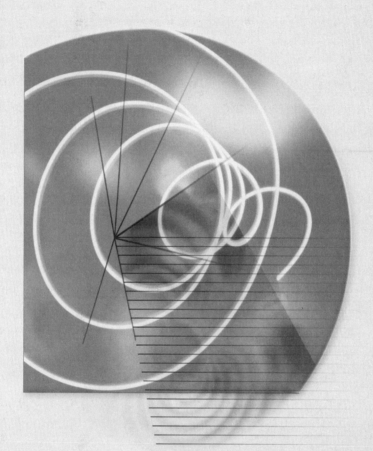

Visual LISP™: A Guide to Artful Programming

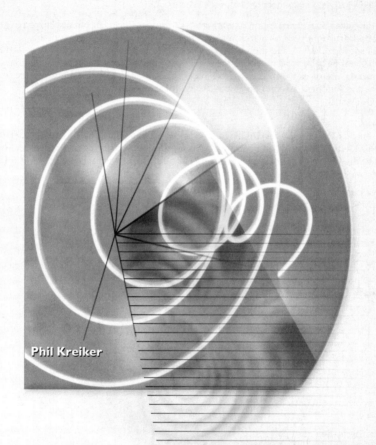

Phil Kreiker

Visual LISP™: A Guide to Artful Programming

Autodesk.
Press

Thomson Learning™

Africa • Australia • Canada • Denmark • Japan • Mexico • New Zealand
Philippines • Puerto Rico • Singapore • Spain • United Kingdom • United States

NOTICE TO THE READER

Trademarks

Autodesk, AutoCAD and the AutoCAD logo are registered trademarks of Autodesk, Inc. Thomson Learning uses "Autodesk Press" with permission from Autodesk, Inc. for certain purposes. Windows is a trademark of the Microsoft Corporation. All other product names are acknowledged as trademarks of their respective owners.

Autodesk Press Staff

Executive Director: Alar Elken
Executive Editor: Sandy Clark
Acquisitions Editor: Michael Kopf
Developmental Editor: John Fisher
Executive Marketing Manager: Maura Theriault
Executive Production Manager: Mary Ellen Black
Production Coordinators: Jennifer Gaines and Larry Main
Art and Design Coordinator: Mary Beth Vought
Marketing Coordinator: Paula Collins
Technology Project Manager: Tom Smith

Cover illustration by Brucie Rosch

Printed in Canada
1 2 3 4 5 6 7 8 9 10 XXX 05 04 03 02 01 00

For more information, contact:
Autodesk Press

3 Columbia Circle, Box 15-015
Albany, New York USA 12212-15015;
or find us on the World Wide Web at http://www.autodeskpress.com

Library of Congress Cataloging-in-Publication Data
Kreiker, Phil.
    Visual LISP: a guide to artful programming/Phil Kreiker.
       p. cm.
    ISBN 0-7668-1549-8
    1. Visual LISP (Computer program language) I. Title.

    QA76.73.V57 K74 2000
    005.13'3—dc21
                                 99-087037

CONTENTS

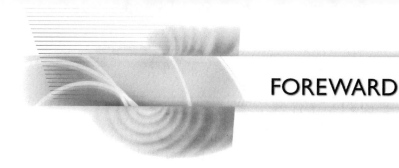

FOREWARD

A WORD FROM THOMSON LEARNING™

Autodesk Press was formed in 1995 as a global strategic alliance between Thomson Learning and Autodesk, Inc. We are pleased to bring you the premier publishing list of Autodesk student software and learning and training materials to support the Autodesk family of products. AutoCAD® is such a powerful product that everyone who uses it would benefit from a mentor to help them unlock its full potential. This is the premise upon which the Programmer Series was conceived. The titles in this series cover the most advanced topics that will help you maximize AutoCAD. Our Programmer Series titles also bring you the best and the brightest authors in the AutoCAD community. Maybe you've read their columns in a CAD journal, maybe you've heard them speak at an Autodesk event, or maybe you're new to these authors—whatever the case may be, we know you'll enjoy and apply what you'll learn from them. We thank you for selecting this title and wish you well on your programming journey.

Sandy Clark
Executive Editor
Autodesk Press

A WORD FROM AUTODESK, INC.

From the birth of AutoCAD onward, there has been a large library of source material on how to use the software; however, relatively little material on customization of AutoCAD has been made available. There have been a few texts written about AutoLISP®, and many general AutoCAD books include a chapter or two on customization. On the whole, however, the number of texts on programming AutoCAD has been inadequate given the amount of open technology, the number of application programming interfaces (APIs), and the sheer volume of opportunity to program the AutoCAD design system.

Four years ago, when I started working with developers as the product manager for AutoCAD APIs, I immediately found an enormous demand for supporting texts about ObjectARX™, AutoLISP, Visual Basic for Applications®, and AutoCAD® OEM. The demand for a "programming series" of books stems from AutoCAD's history of being the most programmable, customizable, and extensible design system on the market. By one measure, over 70 percent of the AutoCAD customer base "programs" AutoCAD using Visual LISP, VBA, or menu customization, all in an effort to increase productivity. The demand for increasing the designer's productivity extends deeply, creating a demand for new source texts to increase productivity of the programming and customization process itself.

The standard Autodesk documentation for AutoCAD provides the original source of technical material on the AutoCAD API technology. However, it tends toward the clinical, which is a natural result of describing the software while it is being created in the software development lab. Documentation in fully applied depth and breadth is only completed through the collective experiences of hundreds of thousands of developers, customers, and users as they interpret and apply the system in ways specific to their needs.

As a result, demand is high for other interpretations of how to use AutoCAD APIs. This kind of instruction develops the technique required to innovate. It develops programmer instinct by instructing when to use one interface over another and provides direction for interpretive nuances that can only be developed through experience. AutoCAD customers and developers look for shortcuts to learning and for alternative reference material. Customers and developers alike want to accelerate their programming learning experience, thereby shortening the time needed to become expert and enabling them to focus sooner and better on their own specific customization or development projects. Completing a customization or development project sooner, faster, and better means greater productivity during the development project and more rapid deployment of the result.

For the CAD manager, increasing productivity through accelerated learning means increasing his or her CAD department's productivity. For the professional developer, this means bringing applications to market faster and remaining competitive.

THE INITIAL IDEA

An incident that occurred at Autodesk University in Los Angeles in the fall of 1997 illustrated dramatically a dynamic demand for AutoCAD API technology information. Bill Kramer was presenting an overview session on AutoCAD's ObjectARX. The Autodesk University officials had planned on having 20 to 30 registrants for this session and had assigned an appropriately sized room to the session. About three weeks before it took place, I received a telephone call indicating that the registrations for the session had reached the capacity of the room, and we would be moving the session. We moved the session two more times due to increasing registration from customers, CAD managers, designers, corporate design managers, and even developers attending the Autodesk University customer event. What they all had in common was an interest in seeing how ObjectARX, an AutoCAD API, was going to increase their own or their department's design productivity. When the presentation started, I walked into the back of the room to see what appeared to me to be over 250 people in a room which was now standing room only. That may have been the moment when I decided to act, or it may have been just when the actual intensity of this demand became apparent to me; I'm not sure which.

The audience was eager to hear what Bill Kramer was going to say about the power of ObjectARX. Nearly an hour of unexpected follow-up questions and answers followed. Bill had successfully evangelized a technical subject to an audience spanning nontechnical to technical individuals. This was a revelation for me and the beginning of my interest in developing new ways to communicate to more people the technical aspects, power, and benefits of AutoCAD technology and APIs.

As a result, I asked Bill if he would write a book on the very topic he just presented. I'm happy to say that Bill's book has become one of the first books written in the new AutoCAD Programmer Series with Autodesk Press.

TOPICS FOR THE PROGRAMMER SERIES

The extent of AutoCAD's open programmable design system made the need for a series of books apparent early on. My team, under the leadership of Cynde Hargrave, senior AutoCAD marketing manager, began working with Autodesk Press to develop this series.

It was an exciting project, with no shortage of interested authors covering a range of topics from AutoCAD's open kernel in ObjectARX to the Windows standard for application programming in VBA. The result is a complete library of references in Autodesk's Programmer Series covering ObjectARX, Visual LISP, AutoLISP, customizing AutoCAD through ActiveX Automation® and Microsoft Visual Basic for Applications, AutoCAD database connectivity, and general customization of AutoCAD.

WHO READS PROGRAMMING BOOKS?

Every AutoCAD user will find books in this series to fit their AutoCAD customization or development interests. The collective goal we had with our team and Autodesk Press for developing this series, identifying titles, and matching them with authors was to provide a broad spectrum of coverage across a wide variety of customization content and a wide range of reader interest and experience.

Collaborating as a team, Autodesk Press and Autodesk developed a programmer series covering all the important APIs and customization topics. In addition, the series provides information that spans use and experience levels from the novice just starting to customize AutoCAD to the professional programmer or developer looking for another interpretive reference to increase his or her experience in developing powerful applications for AutoCAD.

THIRST, THEME, AND VARIATION

I compare this thirst for knowledge with the interest musicians have in listening to music performed in different ways. For me, it is to hear Vivaldi's *The Four Seasons* time and time again. Musicians play from the same notes written on the page, with the identical crescendos and decrescendos and other instructions describing the "technical" aspects of the music.

All of the information to play the piece is there. However, the true creative design and beauty only manifests through the collection of individual musicians, each applying a unique experience and interpretation based on all that he or she has learned before from other mentors in addition to his or her own practice in playing the written notes.

By learning from other interpretations of technically identical music, musicians benefit the most from a new, and unique, interpretation and individual perception. This makes it possible for musicians to amplify their own experience with the technical content in the music. The result is another unique understanding and personalized interpretation of the music.

Similar to musical interpretation is the learning, mentoring, creative processes, and resources required in developing great software, programming AutoCAD applications, and customizing the AutoCAD design system. This process results in books such as Autodesk's Programmer Series, written by industry and AutoCAD experts who truly love working with AutoCAD and personalizing their work through development and customization experience. These authors, through this programmer series, evangelize others, enabling them to gain from their own experiences. For us, the readers, we gain the benefit from their interpretation, and obtain the value through different presentation of the technical information, by this wide spectrum of authors.

Andrew Stein
Senior Manager
Autodesk Business Research, Analysis and Planning

A WORD FROM BILL KRAMER

Caution: Proper use of this book will make you an expert AutoCAD applications programmer.

Although the practice of apprenticeship is not found in the world of computer programming, there are times when it might be nice, especially if you are just learning how to program or have reached one of those tricky places where it is helpful to consult with someone who has been there before. At those times, it is a blessing to have access to the right person or resources to learn how to proceed. Along the same thought, a lot of effort, time, and money can be saved when you are shown how to do something right the first time. And if that information can be imparted in a humorous manner, that is a pleasure to read or hear, it makes the journey to being a master all that much better.

So what does one do in computer programming when seeking guidance to solve a problem? You can attend classes, read books, hire a consultant, or look for examples in magazines and on the Internet. In those cases you hope to find the right expertise, learn from it, and then solve the problem. But sometimes the knowledge required to solve the problem can be quite extensive and only learned from experience. And it is easy to get lost in the details. That is why apprenticeships used to take such a long time in the skilled crafts like metal and wood working, just to name a couple.

When an apprentice starts to learn a craft, they are often not aware of the complexities and subtleties that may be involved. As time goes forward, the apprentice learns things from the master in a deliberate and planned path until they possess the same knowledge and skills. Shortcuts, approaches to standard problems, and the common sense of the trade are all learned at the same time. Once they are under control, the apprentice is then ready to step in the "real world" to practice the craft. The knowledge learned from the master can then be applied in new ways, as the apprentice becomes a master.

In today's world of computer applications development, we do not have apprenticeships, but we do have masters of the craft. Some of them even have the ability to share that experience and expertise in fun ways. Phil Kreiker is one of them. His approach to programming is wonderful and enjoyable all at once. He has provided a tool kit in this book that will enable an apprentice to do fantastic things as well as provide other masters with insights into how things should be done.

The "Artful Programming Application Programming Interface" (AP-API) presented in this book is more than just a set of utilities. It is a philosophy of how to do it right the first time. All levels of programmers will gain knowledge and tips as they work with these contents. At the same time, it is presented in Phil's unique way of making the challenge of programming applications fun. After reviewing a pre-release of the book I even went back to look at some of my previous work to see how I could improve it.

Phil does more than present a basic how-to in this book. He imparts knowledge gained through years of experience as a master AutoCAD applications programmer—little things like making your users happy with the software you write and making your applications look like AutoCAD itself.

He also provides patches to some of the more frustrating points in the Visual LISP programming language, making it easier to get the job done in the manner best suited to the job. For example, the AP-API simplifies the use of ActiveX components and interfacing with the Visual Basic for Applications (VBA) provided in AutoCAD, as well as the Microsoft Office Suite. One thing that has frustrated even experienced programmers is the need to switch entity data types when using the ActiveX functions of Visual LISP. The AP-API opens the door to those functions and does so in a manner more easily adapted by experienced AutoLISP programmers wanting to take advantage of the new tools, but who are put off by the seemingly complex requirements of talking to VBA and Object ARX.

A great deal of experience can be found in the pages of this book, and even greater treasures await the programmer who explores the CD in detail. Plus, it is fun to hear Phil's comments along the way as he bounds from one topic to the next with the dexterity of the rabbit hopping through a garden.

By reading and using the information in this book, you will shave years of experimentation off your journey to becoming a master applications programmer. Virtually any systems utility you can think of will be found inside, and there are a lot more awaiting your discovery. This book and the utilities provided are a must for programmers who write applications every so often and want to have a professional, polished look to their works. As for expert programmers, you will enjoy learning new tricks and techniques as you dig deeper into the content. Only a true master like Phil can make a book and utility set like this one.

I have had the pleasure of knowing Phil for a number of years and have always enjoyed his style, wit, and profound understanding of the problems that face CAD/CAM/CAE programmers. We both have taught other programmers, as well as each other, a lot about the philosophy of programming computers for AutoCAD applications. If you don't already know this, engineers and architects can be a demanding lot when it comes to applications. Not only do they expect it to be technically correct at all times, but they also don't want the software to get in the way of what they do. And *that* is where the creativity really comes in to play.

You will enjoy this book, and you will learn from it, but I must caution you—after reading this book and using the utilities provided on the accompanying CD, you will not look at AutoCAD applications the same way any more. You will see them from the eyes of a master, as you will have entered a new way of getting the job done right. If you feel like Alice in Wonderland while walking with Phil, don't be too surprised. His columns titled "Down the Rabbit Hole" that now appear at his Web site, (www.lgmicro.com), have that name for a reason. So sit back, and prepare to learn the Artful Programmer's way of doing things.

Keep on programmin'.

Bill Kramer
AUTO CODE Software

The Bridge Builder
by Will Allen Dromgoole

An old man, going a lone highway,
Came at the evening, cold and gray,
To a chasm vast and deep and wide;
The old man crossed in the twilight dim,
The sullen stream had no fear for him;
But he turned when safe on the other side,
And built a bridge to span the tide.

"Old man," said a fellow pilgrim near,
"You are wasting your strength in building here;
Your journey will end at the ending day,
You never again will pass this way;
You have crossed the chasm deep and wide;
Why build you this bridge at evening tide?"

The builder lifted his old gray head—
"Good friend, in the path I have come," he said,
"There followeth after me today,
A youth whose feet must pass this way;
This chasm that was naught to me
To that fair-haired youth may a pitfall be;
He, too, must cross in the twilight dim—
Good friend, I am building this bridge for him."

PREFACE

WHAT IS ARTFUL?

For your edification, enlightenment, and entertainment, I've pulled these definitions from Microsoft *Bookshelf'99*:

art·ful

art·ful (art'ful) *adjective*

1. Exhibiting art or skill: "The furniture is an artful blend of antiques and reproductions" (Michael W. Robbins).

2. Skillful in accomplishing a purpose, especially by the use of cunning or craft. See synonyms at *sly*.

3. Artificial.

— **art'ful·ly** *adverb*

— **art'ful·ness** *noun*[1]

WHAT IS ARTFUL PROGRAMMING?

Let us define "artful programming" as *the artful creation of artful programs.*

You may feel free to use any or all of the above meanings of *artful* in this definition. I'm willing to concede artful programming as artificial, only insofar as all human endeavors are deemed (by some) to be artificial.

Artful programs help users meet their goals.

According to Alan Cooper, there are three very good reasons for writing programs that help users meet their goals:

1. If we write programs that help users meet their goals, they will be happy.

2. If our programs make people happy, they will buy our programs.

3. If people buy our programs, we will be rich, famous, and happy.

Even if *you* are the only user of a program you are writing, if you write programs that help you meet *your* goals, then *you* will be happy. If you're writing programs that *don't* help you meet your goals, then why are you writing them?

ARTFUL PROGRAMS NEVER DO ANYTHING THEIR USERS DO NOT REASONABLY EXPECT THEM TO DO

How can we possibly help people meet their goals with tools that behave *unreasonably*?

Artful programs never invoke the Blue Screen of Death. Artful programs never inadvertently reformat your hard drive. Artful programs never spit in your beer.

ARTFUL PROGRAMS DO NOT MAKE YOUR COMPUTER SCHIZOPHRENIC

Joel Orr first pointed out to me the schizophrenic nature of computer systems. "If one day," he said to me, "you were say 'Hello, Joel,' and I responded 'Hi, Phil,' and on another day, you say 'Hello, Joel,' and I respond with a punch in your mouth, I would be exhibiting multiple personalities; I would be schizophrenic. If your computer programs exhibit multiple personalities, your computer would be schizophrenic."

One of the first PC-based word processing programs I used was WordStar. It was hard to learn, but easy to use. The key sequence for search and replace was CTRL+Q,A (as in Quick Alter).

WordStar 4 begot WordStar 2000. This new, improved WordStar used CTRL+Q,A to Quit and Abandon changes. If this isn't schizophrenia, I don't know what is. (I quit and abandoned WordStar 2000.)

ARTFUL PROGRAMS BEHAVE AS DO AUTOCAD'S INTRINSIC COMMANDS

Let's face it. If you're adding commands to AutoCAD, users reasonably expect them to behave (at least somewhat) like AutoCAD's intrinsic commands. Thus, a STAR command should have about the same prompts as the POLYGON command.

IT ISN'T DIFFICULT TO WRITE ARTFUL PROGRAMS

It just takes an earnest desire to do so. The Artful Programming API (Application Programming Interface) makes it easy.

WHO SHOULD READ THIS BOOK?

This book is written for the intermediate to advanced AutoLISP programmer. It is assumed that you have taken the Visual LISP tutorial that ships with AutoCAD 2000.

WHAT'S ON THE ARTFUL PROGRAMMING CD-ROM?

- The Artful Programming API Online Reference and Program Listings—*ap-api.chm*
- The Artful Programming API—*ap-api.fas*
- The HTML Help Update Program—*hhupd.exe*

CAN I DISTRIBUTE APPLICATIONS BASED ON THE ARTFUL PROGRAMMING API ROYALTY-FREE?

You bet your sweet bippie you can. Just four rules:

1. You may not distribute the Artful Programming API (*ap-api.fas*) itself.

2. You may distribute any applications you create with the Artful Programming API as long as they are Compiled (.vlx) Separate Namespace Applications. See Chapter 7 for detailed instructions.

3. You must distribute the unmodified Artful Programming API Redistribution Archive in its entirety.

4. You may neither remove nor disable nor disguise our copyright notice.

LEGAL MATTERS

LOOKING GLASS MICROPRODUCTS TRADEMARKS

The following are trademarks of Looking Glass Microproducts, Inc, in the USA and/or other countries: Phil Kreiker's Underware, Computer Underware, Underware for Computers.

AUTODESK TRADEMARKS

The following are registered trademarks of Autodesk, Inc., in the USA and/or other countries: 3D Plan, 3D Props, 3D Studio, 3D Studio MAX, 3D Studio VIZ, 3D Surfer, ADE, ADI, Advanced Modeling Extension, AEC Authority (logo), AEC-X, AME, Animator Pro, Animator Studio, ATC, AUGI, AutoCAD, AutoCAD Data Extension, AutoCAD Development System, AutoCAD LT, AutoCAD Map, Autodesk, Autodesk Animator, Autodesk (logo), Autodesk MapGuide, Autodesk University, Autodesk View, Autodesk WalkThrough, Autodesk World, AutoLISP, AutoShade, AutoSketch, AutoSolid, AutoSurf, AutoVision, Biped, bringing information down to earth, CAD Overlay, Character Studio, Design Companion, Drafix, Education by Design, Generic, Generic 3D Drafting, Generic CADD, Generic Software, Geodyssey, Heidi , HOOPS, Hyperwire, Inside Track, Kinetix, MaterialSpec, Mechanical Desktop, Multimedia Explorer, NAAUG, Office Series, Opus, PeopleTracker, Physique, Planix, Rastation, Softdesk, Softdesk (logo), Solution 3000, Tech Talk, Texture Universe, The AEC Authority, The Auto Architect, TinkerTech, WHIP!, WHIP! (logo), Woodbourne, WorkCenter, and World-Creating Toolkit.

The following are trademarks of Autodesk, Inc., in the USA and/or other countries: 3D on the PC, ACAD, ActiveShapes, Actrix, Advanced User Interface, AEC Office, AME Link, Animation Partner, Animation Player, Animation Pro Player, A Studio in Every Computer, ATLAST, Auto-Architect, AutoCAD Architectural Desktop, AutoCAD Architectural Desktop Learning Assistance, AutoCAD DesignCenter, Learning Assistance, AutoCAD LT Learning Assistance, AutoCAD Simulator, AutoCAD SQL Extension, AutoCAD SQL Interface, AutoCDM, Autodesk Animator Clips, Autodesk Animator Theatre, Autodesk Device Interface, Autodesk PhotoEDIT, Autodesk Software Developer's Kit, Autodesk View DwgX, AutoEDM, AutoFlix, AutoLathe, AutoSnap, AutoTrack, Built with ObjectARX (logo), ClearScale, Concept Studio, Content Explorer, cornerStone Toolkit, Dancing Baby (image), Design Your World, Design Your World (logo), Designer's Toolkit, DWG Linking, DWG Unplugged, DXF, Exegis, FLI, FLIC, GDX Driver, Generic 3D, Heads-up Design, Home Series, Kinetix (logo), MAX DWG, ObjectARX, ObjectDBX, Ooga-Chaka, Photo Landscape, Photoscape, Plugs and Sockets, PolarSnap, Powered with Autodesk Technology, Powered with Autodesk Technology (logo), ProConnect, ProjectPoint, Pro Landscape, QuickCAD, RadioRay, SchoolBox, SketchTools, Suddenly Everything Clicks, Supportdesk, The Dancing Baby, Transforms Ideas Into Reality, Visual LISP, and Volo.

THIRD-PARTY TRADEMARKS

Microsoft, Visual Basic, Visual C++, and Windows are registered trademarks and Visual FoxPro and the Microsoft Visual Basic Technology logo are trademarks of Microsoft Corporation in the United States and other countries.

All other brand names, product names, or trademarks belong to their respective holders.

WHAT ABOUT UPDATES/UPGRADES?

Updates to the Artful Programming API are available at no charge from our web site, www.ComputerUnderware.com. I recommend that you download the latest version of the software.

WHEN YOU'RE DONE WITH THIS BOOK

You'll know how to make AutoCAD help you and your users meet their goals. That should make them happy, which, as we've already pointed out, will make you rich, famous, and happy. More than that, I cannot promise.

TYPOGRAPHICAL CONVENTIONS

For the sake of clarity, specific items are set in typefaces to distinguish them from body text.

For the sake of consistency, we'll be following (but not too closely) the typographical conventions of the *AutoCAD 2000 Visual LISP Developer's Guide* and the *AutoCAD 2000 Customization Guide*:

Text Element	Example
Program code examples are shown in monospaced type.	`(defun mad-hatter ( )` `(setq *MARCH-HARE* 4.0))`
Prompts are shown in monospaced type.	`Select object to trim or [Project/` `Edge/Undo]:`
Instructions after prompt sequences are shown in italics and enclosed in parentheses.	`Select objects:` *(Use an object selection method)*
Text you enter is shown in boldface.	At the VLISP Console prompt, enter **laughing**
Keys you press on the keyboard.	CTRL, F12, ESC, ENTER
Keys you press simultaneously on the keyboard.	CTRL+C
File names and folder names are shown in italics when referred to in a sentence.	Double-click the file name *doormouse.lsp*. The default install directory is *C:\Program Files\ACAD2000*

Text Element	Example
In references that include variable text, the variable text is in italics.	A FAS file named *appname*-init.fas, where *appname* is the application module name
AutoLISP variable names are shown in monospaced type.	Double-click on any occurrence of the variable name `caterpillar` The `*MAD-HATTER*` system variable
AutoLISP function names are shown in bold-monospaced type.	Use the **rem** function... Use **setenv** to set
Formal arguments specified in function definitions are shown in italic-monospaced type.	The *string* and *mode* arguments...
AutoCAD commands are displayed in small caps.	ADCENTER, DBCONNECT, SAVE
AutoCAD system variables are displayed in small caps.	DIMBLK, DWGNAME, LTSCALE
AutoCAD named objects, such as line types and styles, are displayed in caps.	DASHDOT, STANDARD

A MATTER OF STYLE

Bill Kramer's comments on style were well appreciated; I'd like to share them with you.

Bill: The variable name l (elle) looks like the digit 1 (won) in the type font selected.

Phil: I replaced the variable name l (elle) with the variable n (en). It's much more legible.

Bill: The symbol ENAME is a return value from the type function. I recommend you not use ename as variable name, to avoid confusion.

Phil: I changed them all to e-name.

Bill: The use of **setq** is not consistent. Sometimes **setq** is used in a sequence of assignments, other times it is used for each assignment, and in others, it is mixed. I often am asked about this in *my* program examples, perhaps you should comment on this.

Phil: It was Ralph Waldo Emerson who said, "A foolish consistency is the hobgoblin of little minds... With consistency a great soul has simply nothing to do."[2]

That said, I feel compelled to explain some of the styles I've used in this book.

---

[2] *Encarta® Book of Quotations* © & (P) 1999 Microsoft Corporation. All rights reserved. Developed for Microsoft by Bloomsbury Publishing Plc.

Why do I use

```
(vl-cmdf "_circle" p0 1.0)          ; draw the first circle
(vl-cmdf "_circle" p0 0.9)          ; draw the second
```

when

```
(vl-cmdf "_circle" p0 1.0          ; draw the first circle
  "_circle" p0 0.9)                 ; draw the second
```

would work?

By having each call to **vl-cmdf** or **command** represent a complete interaction with AutoCAD from Command: prompt to Command: prompt,

- it's easier for me to remember what I was trying to do.

- it's easier for me to insert, delete, edit, and debug the command sequences.

- the performance should be about the same, thanks to Visual LISP's optimizing compiler.

Why do I use

```
setq size 0.5) ; size of cross
setq p1 (ap-vector-dif p0 (list size 0 0))     ; left side
p2 (ap-vector-sum p0 (list size 0 0))          ; right side
p3 (ap-vector-dif p0 (list 0 size 0))          ; bottom
p4 (ap-vector-sum p0 (list 0 size 0))          ; top
)
```

when

```
(setq size 0.5 ; size of cross
p1 (ap-vector-dif p0 (list size 0 0))          ; left side
p2 (ap-vector-sum p0 (list size 0 0))          ; right side
p3 (ap-vector-dif p0 (list 0 size 0))          ; bottom
p4 (ap-vector-sum p0 (list 0 size 0))          ; top
)
```

or

```
(setq size 0.5) ; size of cross
(setq p1 (ap-vector-dif p0 (list size 0 0)))   ; left side
(setq p2 (ap-vector-sum p0 (list size 0 0)))   ; right side
(setq p3 (ap-vector-dif p0 (list 0 size 0)))   ; bottom
(setq p4 (ap-vector-sum p0 (list 0 size 0)))   ; top
```

would work?

By having each call to **setq** represent a logical assignment group,

- it's easier for me to remember what I was trying to do.
- it's easier for me to insert, delete, edit, and debug the function sequences.
- the performance should be about the same, thanks to Visual LISP's optimizing compiler.
- it's easier to avoid such mistakes as replacing

```
(if expression
    (setq a 0)
    (setq b 1)
)
```

with

```
(if expression
    (setq a 0
          b 1)
)
```

ABOUT THE AUTHOR

With a master of science degree in Electrical Engineering, Phil Kreiker has been involved in many facets of the computer industry since 1963: hardware, software, human factors, systems design, and production control.

Phil has been writing and using CAD systems since 1968. Using these systems, he has programmed and designed computers and peripherals for Data General Corporation, Digital Equipment Corporation, Datatrol, and Raytheon Company. He's implemented and managed systems for the United Nations International Computing Centre and for Hewlett-Packard.

He has taught AutoCAD to design conveyor systems, lay bricks, design shock-absorber boots, convert orthographic projections into isometric ones, string transmission lines, cast shadows, map mines, and more.

Phil has also found time to master a plethora of computer languages, author numerous articles and books, and receive a patent for a microprocessor controller circuit.

He teaches AutoCAD at the Colorado School of Mines, and was technical editor at large at *CADENCE Magazine* from 1996–1997. Phil was a contributing editor at *CADalyst Magazine* from 1988 to 1996. His first book, *The CAD Cookbook Collection*, was published in 1993.

In 1990, 1992, and again in 1998, Phil came in second in the AutoCAD Top Gun U.S.A. competition at AutoCAD Expo conferences; he was elected to the board of directors of the North American AutoCAD User Group in 1990. In 1991, Phil was elected president of the North American AutoCAD User Group. In 1992, he assumed the position of president of the user group.

In 2000, Phil came up with the concept of Phil Kreiker's Underware for Computers.

Phil, his wife, Joy, and their two children moved to Loveland, Colorado, in 1979. Joy and Phil Kreiker established Looking Glass Microproducts in 1982; it was named Business of the Year in 1990 by the Colorado Association of Commerce and Industry.

ACKNOWLEDGMENTS

This book would not have been possible without the support, trust, and encouragement of Joy Blauvelt Kreiker of Looking Glass Microproducts, Lynn Allen of Autodesk, and Allison Powell and John Fisher of Autodesk Press. Thank you all.

Special thanks must go to Bill Kramer for the technical review of the manuscript, his comments on style, and his writing the foreword to this book. Bill has promised me he won't deny that it's *exactly* as he submitted it to me, aside from a few gratuitous commas I added, just to give the reader a place to rest their eyes. Thanks, Bill.

DEDICATION

I've dedicated this book to a number of people.

First, it is dedicated, for reasons known only to her and me, to my muse, whom some call Joy Blauvelt Kreiker. Joy secretly wishes to be called Tina Sparkle.

Next, it is dedicated to my first daughter, whom some call Pepper Gagliano Forrest. I don't know what she secretly wishes to be called. Pepper is the nicest person I know and has taught me, above all, to be nice to others, *but only as much as one can.*

Third, it is dedicated to my second daughter, whom some call China Blauvelt Kreiker. Sometimes, she has wished to be called Crystal. I don't know how she wishes it to be spelled. China is the most fascinating person I know, and is currently reforesting El Salvador with the Peace Corps.

In addition to these three, this book is dedicated to the health and memory of every woman I've ever known. To those of you whom I have offended by word or deed, I submit that I had mistaken you for someone else. To those of you whom I have offended by not being there when you needed me, I submit that I forgot.

Phil Kreiker
June, 2000
Loveland, Colorado

CHAPTER 1

Installing the Artful Programming API

The Artful Programming API is a collection of AutoLISP functions that greatly facilitate the development and debugging of artful AutoLISP programs. If you wish to use the Artful Programming API on your computer, you must first set it up.

Updates to the Artful Programming API are available at no charge from our web site, www.ComputerUnderware.com. I recommend that you download the latest version of the Artful Programming API.

SETUP

Just run *setup.exe* on the Artful Programming CD-ROM. This will install the Artful Programming API files on your system, along with the online documentation.

CONFIGURING AUTOCAD

If you're going to be developing applications with the Artful Programming API, I recommend that you add the file *ap-api.fas* to your startup suite. After that, it will be loaded automatically each time you open (or create) a drawing. Here's how:

First, unzip the file *ap-api.fas* from the Artful Programming CD-ROM to your hard drive. It doesn't matter where you copy it, just remember where it is.

Launch AutoCAD 2000, and Select Tools|Load Application...

This will display the Load/Unload Applications dialog box, shown in Figure 1–1.

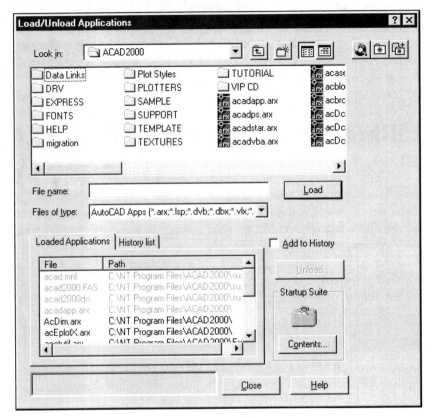

Figure 1–1

Under Startup Suite, click on the Contents... button. This will display the Startup Suite dialog box, shown in Figure 1–2.

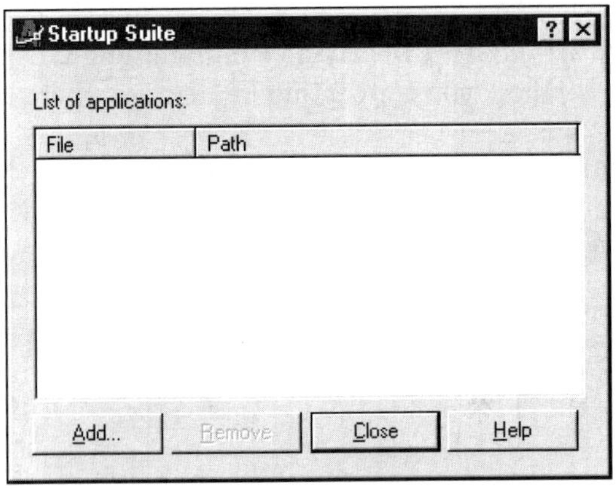

Figure 1–2

Click on the Add... button. This will display the Add File to Startup Suite dialog box, as shown in Figure 1–3.

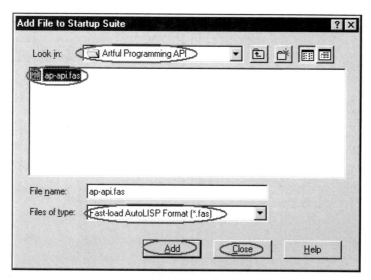

Figure 1–3

From here, navigate to your *ap-api.fas* file (you do remember where you put it, don't you?), then click on the Add button, then on the Close button. This will return you to the Startup Suite dialog box, as shown in Figure 1–4.

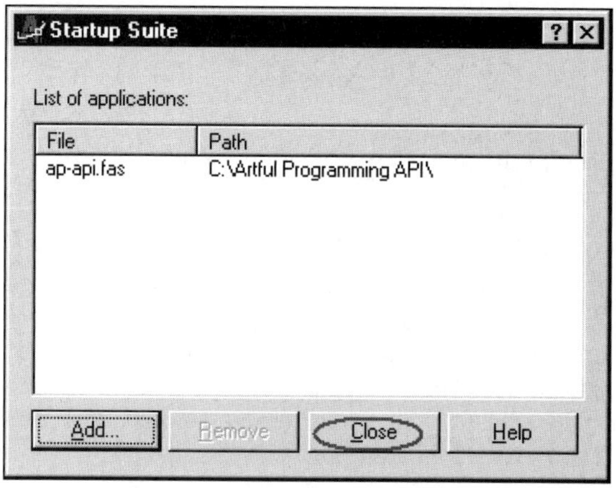

Figure 1–4

Click on Close. This will return you to the Load/Unload Applications dialog box as shown in Figure 1–5.

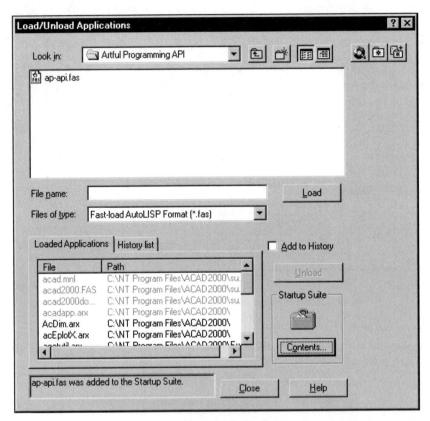

Figure I–5

Note the message to the left of the Close button. Click on Close.

Henceforth, *ap-api.fas* will be loaded automatically each time you open (or create) a drawing. You should see a message as follows at the command prompt:

```
ap-api.fas was added to the Startup Suite. Command:
   Powered by Phil Kreiker's Underware -- The Artful
   Programming API v15.0.0, © 2000 by Looking Glass
   Microproducts, Inc.
```

CHAPTER 2

Manipulating the AutoCAD User

This chapter shows how the Artful Programming API makes it easier to manipulate the AutoCAD user.

THE PRIME DIRECTIVE

The best way I've discovered to write artful programs that help users meet their goals (to make them happy, have them buy our software, and make us lots of money) is to follow the prime directive:

 Prime Directive: No program should ever do anything a user does not reasonably expect it to do.

This, of course, begs the question: "What, then, do users reasonably expect?"

Given that our users are presumably working in an AutoCAD environment, it's safe to assume that they expect our commands to look and feel like AutoCAD's intrinsic (built-in) commands. Let us then restate the prime directive as follows:

 Artful Prime Directive: No AutoCAD program should ever do anything an AutoCAD user does not reasonably expect it to do.

This, in turn, begs the question: "How, then, do we emulate AutoCAD's intrinsic commands?"

The method we'll use here is what I call 'sculpting the elephant.' You start with a block of marble and remove everything that doesn't look like an elephant. In AutoCAD terms, it means we'll wind up with a template to be used for all of the commands we create.

Let's start with the program as shown in Listing. 2–1.

Listing 2–1

```
;;; Filename: AP-02-01.lsp
(defun c:dcircle (/ p0)
  (setq p0                                  ; center of circles
        (getpoint "\nSpecify center point for circles: ")
  )
  (vl-cmdf "_circle" p0 1.0)               ; draw the first circle
  (vl-cmdf "_circle" p0 0.9)               ; draw the second
)
```

 Tip: All listings may be found in the file *ap-api.chp* in *setup.exe* on the Artful Programming CD-ROM. You may paste these listings into the Visual LISP Editor window.

This command should draw a pair of circles at a user-specified point. Let's give it a try.

Select Tools|Load Text in Editor

Switch back to AutoCAD and enter the following at the Command: prompt:

```
Command: cmdecho
Enter new value for CMDECHO <default>: 1 (Sets cmdecho to 1.)
Command: blipmode
Enter mode [ON/OFF] <default>: on (Sets blipmode on.)
Command: dcircle (Draws two circles.)
Specify center point for circles: (Specify center point.)
_circle Specify center point for circle or [3P/2P/Ttr (tan tan
    radius)]: Specify radius of circle or [Diameter] <0.9000>:
    1.000000000000000
Command: _circle Specify center point for circle or [3P/2P/Ttr
    (tan tan radius)]: Specify radius of circle or [Diameter]
    <1.0000>: 0.900000000000000
Command: T
Command:
```

QUIETING THINGS DOWN

By default, AutoCAD echoes the results of the **command** and **vl-cmdf** functions as though you had typed the commands at the Command: prompt. Setting the CMDECHO system variable to 0 turns this off.

What's that T all about? Every AutoLISP function returns a value. In this case, the **c:dcircle** function returns the value of the **vl-cmdf** function, which is always T. Visual LISP prints this value at the Command: prompt.

Because the user does not reasonably expect to see T, we need to return some other value, ideally one that does not print anything.

As it turns out, there *is* a function that returns a nonprinting value. It's the function known as **princ**. End your programs with **(princ)**, and say goodbye to T.

Return to Visual LISP and edit (or load) the program shown in Listing 2–2.

Listing 2–2

```
;;; Filename: AP-02-02.lsp
(defun c:dcircle (/ p0)
  (setvar "cmdecho" 0)                    ; quiet things down
  (setq p0                               ; center of circles
       (getpoint "\nSpecify center point for circles: ")
  )
  (vl-cmdf "_circle" p0 1.0)             ; draw the first circle
  (vl-cmdf "_circle" p0 0.9)             ; draw the second
  (princ)                                ; exit quietly
)
```

Select Tools|Load Text in Editor

Switch to AutoCAD and type the following at the Command: prompt:

```
Command: dcircle
Specify center point for circles: (SpecifySpecify center point.)
Command: (Draws two circles.)
```

That's nicer now, isn't it? The time you spend inserting that **(princ)** statement into your commands will certainly be less that the time you spend saying (even once), "Oh, 'T', that's perfectly normal. It means the command is done."

Is this all that's wrong with this command? Not by a long shot.

SAVING AND RESTORING SYSTEM VARIABLES

Switch back to AutoCAD, and type the following at the Command: prompt:

```
Command: circle  (Draws a 2-unit circle.)
Specify center point for circle or [3P/2P/Ttr (tan tan radius)]:
    (Specify center point.)
Specify radius of circle or [Diameter] <default>: 2
Command: dcircle
Specify center point for circles: (Specify center point.)
Command: circle  (Draws a 0.9-unit circle.)
Specify center point for circle or [3P/2P/Ttr (tan tan radius)]:
    (Specify center point.)
Specify radius of circle or [Diameter] <0.9000>: ENTER
Command:
```

As an experienced AutoCAD user, you expected AutoCAD to default the circle radius of the second CIRCLE command to the value that *you* supplied for the first CIRCLE command. Unfortunately, the **command** and **vl-cmdf** functions set the default circle radius to the radius of the last circle *they* drew. This is *not* what the user reasonably expects.

Return to Visual LISP and edit (or load) the program shown in Listing 2–3.

Listing 2–3

```
;;; Filename: AP-02-03.lsp
(defun c:dcircle (/ p0 AP:SYSVARS)
  (setvar "cmdecho" 0)                    ; quiet things down
  (ap-push-vars '(("circlerad")))         ; save system variables
  (setq p0                                ; center of circles
        (getpoint "\nSpecify center point for circles: ")
  )
  (vl-cmdf "_circle" p0 1.0)              ; draw the first circle
  (vl-cmdf "_circle" p0 0.9)              ; draw the second
  (ap-pop-vars)                           ; restore system variables
  (princ)                                 ; exit quietly
)
```

Any system variables we need to save before running a command may be saved with the **ap-push-vars** function and restored with the **ap-pop-vars** function.

Tips: Any function whose name begins *ap-* is part of the Artful Programming API.

Before you can access the Artful Programming API, you must install it, as discussed in Chapter 1.

Select Tools|Load Text in Editor.

Switch to AutoCAD, and enter the following at the Command: prompt:

```
Command: circle   (Draws a 2-unit circle.)
Specify center point for circle or [3P/2P/Ttr (tan tan radius)]:
   (Specify center point.)
Specify radius of circle or [Diameter] <default>: 2
Command: dcircle
Specify center point for circles: (Do so; it draws two circles.)
Command: circle   (Draws a 2-unit circle.)
Specify center point for circle or [3P/2P/Ttr (tan tan radius)]:
   (Specify center point.)
Specify radius of circle or [Diameter] <2.0000>: ENTER
Command:
```

Notice that now the default radius for the second CIRCLE command is the radius used in the first. Pretty spiffy, huh?

NICE TO SEE U'S

Let's see how our program deals with the U command.

Enter the following at the AutoCAD Command: prompt:

```
Command: u
CIRCLE  (Removes the last circle.)
Command: u
CIRCLE GROUP:  (Removes 0.9-unit circle.)
Command: u
CIRCLE GROUP:
Command:  (Removes 1.0-unit circle.)
```

Note well that it took *two* U commands to undo the two circles drawn by `dcircle`. As an experienced AutoCAD user, you reasonably expected *both* circles drawn by the `dcircle` command to be undone with a single U.

Making sure that everything you do in a command is undone by a single U or UNDO 1 command is a little more complicated than it seems at first glance.

According to the AutoCAD documentation, you may group operations for U by bracketing the stuff to be undone between

```
(vl-cmdf "_.undo" "_begin")
```

and

```
(vl-cmdf "_.undo" "_end")
```

Ah, if only life were that simple. There are three problems (or opportunities) here for the artful programmer:.

1. If UNDO control has been set to None or to One, the UNDO command will reject the keywords Begin and End. Begin and End are irrelevant in these contexts, so you get an "Invalid option keyword" error, and AutoCAD stays inside the UNDO command.

   So, if you don't want your routines to crash most unexpectedly (after all, almost no one sets UNDO to None or One, but *someone will,* and you don't want to be spending your nights trying to figure out *this* one), you must test for these cases in every AutoCAD command that you write.

   The alternative is to wait for Autodesk to fix the UNDO command, but that could take some time. Instead, we must abstain from issuing the UNDO Begin and UNDO End sequences.

2. If an undo-group is active when you start your command, everything your command does is supposed to be part of the currently active undo-group. This allows AutoCAD to automatically group commands evoked from the menu bar or from buttons.

   Unfortunately, issuing an UNDO Begin or an UNDO End while a group is active will end the active undo-group. We must again abstain from this option.

3. As near as I can tell, AutoCAD gets confused if you have UNDO Auto on between an UNDO Begin and an UNDO End.

The good news is that we need deal with this nonsense only once (per program, that is). Here's how we do it.

Return to Visual LISP and edit (or load) the program shown in Listing 2–4.

Listing 2–4

```
;;; Filename: AP-02-04.lsp
(defun c:dcircle (/ p0 AP:SYSVARS AP:UNDOCTL)
  (setvar "cmdecho" 0)                      ; quiet things down
  (ap-push-undo nil)                        ; start undo-group
  (ap-push-vars '(("circlerad")))           ; save system variables
  (setq p0                                  ; center of circles
        (getpoint "\nSpecify center point for circles: ")
  )
  (vl-cmdf "_circle" p0 1.0)                ; draw the first circle
  (vl-cmdf "_circle" p0 0.9)                ; draw the second
  (ap-pop-vars)                             ; restore system variables
  (ap-pop-undo)                             ; end undo-group
  (princ)                                   ; exit quietly
)
```

Bracketing our program with the **ap-push-undo** and **ap-pop-undo** functions (from the Artful Programming API) ensures that the U command functions as reasonably expected.

Select Tools|Load Text in Editor.

Switch to AutoCAD and enter the following at the Command: prompt:

```
Command: undo
Enter the number of operations to undo or [Auto/Control/BEgin/
   End/Mark/Back]: e (Assures no undo-group is active.)
Command: dcircle
Specify center point for circles: (Specify center point; draws two circles.)
Command: u
GROUP (UNDOs both circles.)
Command: redo
Command: (REDOs both circles.)
```

Are we done? No, not yet.

THERE'S NO ESCAPE

Let's see how our program deals with the ESC key: What does it do when you want to bail out?

```
Command: circle
Specify center point for circle or [3P/2P/Ttr (tan tan radius)]:
    ESC *Cancel* (Cancels with no error message.)
Command: dcircle
Specify center point for circles: ESC *Cancel*
; error: Function cancelled
Command: (Displays unexpected error message.)
```

In addition to the unexpected error message, the default Visual LISP error handler ***error*** does not end our undo-group or restore our system variables as we would expect it to do. Let's fix it.

Return to Visual LISP and edit (or load) the program shown in Listing 2–5.

Listing 2–5

```
;;; Filename: AP-02-05.lsp
(defun c:dcircle (/ p0 AP:SYSVARS AP:UNDOCTL AP:OLD-ERROR)
  (setvar "cmdecho" 0)                      ; quiet things down
  (ap-push-error ap-error)                  ; save error handler
  (ap-push-undo nil)                        ; start undo-group
  (ap-push-vars '(("circlerad")))           ; save system variables
  (setq p0                                  ; center of circles
        (getpoint "\nSpecify center point for circles: ")
  )
  (vl-cmdf "_circle" p0 1.0)                ; draw the first circle
  (vl-cmdf "_circle" p0 0.9)                ; draw the second
  (ap-pop-vars)                             ; restore system variables
  (ap-pop-undo)                             ; end undo-group
  (ap-pop-error)                            ; restore error handler
  (princ)                                   ; exit quietly
)
```

The **ap-push-error** and **ap-pop-error** functions allow us to override the current AutoLISP error handler. This ensures that, in case of an error, the saved system variables, error handler, and undo controls are restored to their prior values.

Select Tools|Load Text in Editor.

Switch to AutoCAD, and enter the following at the Command: prompt:

```
Command: dcircle Specify center point for circles: ESC *Cancel*
Command:
Command:
```

Now that looks better.

 Note: I have yet to find a way to avoid the extra blank Command: prompt. Oh, well.

At least we're looking more and more like an intrinsic AutoCAD command. Are we done? Not quite yet.

TRIPPING UP TRANSPARENTLY

Let's see what happens when we transparently invoke our DCIRCLE command:

Enter the following at the Command: prompt:

```
Command: line
Specify first point: 2,2 (Start the LINE command.)
Specify next point or [Undo]: 'circle  (Evoke CIRCLE transparently.)
Point or option keyword required. (And it's ignored.)
Specify next point or [Undo]: 'dcircle  (Evoke DCIRCLE transparently.)
Specify center point for circles: (Specify center point. It starts running . . . )
Point or option keyword required.
Point or option keyword required.
Point or option keyword required.
Point or option keyword required. ( . . . but fails when it tries to evoke the
    CIRCLE command transparently.)
Zero length line created at (12.2297, 0.8813, 0.0000)
Point or option keyword required.
Specify next point or [Undo]: ENTER
Command: (End the LINE command.)
```

We need to prevent the DCIRCLE command from being invoked transparently.

Return to Visual LISP and edit (or load) the program shown in Listing 2–6.

Listing 2–6

```
;;; Filename: AP-02-06.lsp
(defun c:dcircle (/ p0 AP:SYSVARS AP:UNDOCTL AP:OLD-ERROR)
  (setvar "cmdecho" 0)                    ; quiet things down
  (ap-push-error ap-error)                ; save error handler
  (ap-exit-transparent)                   ; exit if transparently
  (ap-push-undo nil)                      ; start undo-group
  (ap-push-vars '(("circlerad")))         ; save system variables
  (setq p0                                ; center of circles
        (getpoint "\nSpecify center point for circles: ")
  )
  (vl-cmdf "_circle" p0 1.0)              ; draw the first circle
  (vl-cmdf "_circle" p0 0.9)              ; draw the second
  (ap-pop-vars)                           ; restore system variables
  (ap-pop-undo)                           ; end undo-group
  (ap-pop-error)                          ; restore error handler
  (princ)                                 ; exit quietly
)
```

Select Tools|Load Text in Editor.

Switch to AutoCAD and enter the following at the Command: prompt:

```
Command: line
Specify first point: 2,2 (Start the LINE command.)
Specify next point or [Undo]: 'dcircle   (Evoke DCIRCLE transparently.)
** This command may not be invoked transparently **
Specify next point or [Undo]:   (We get an extra prompt here.)
Specify next point or [Undo]: ENTER
Command: (Quit.)
```

That looks better. But are we done? Nope.

Note: I have not yet found a way to avoid the extra prompts when exiting a command that should not be evoked transparently.

THE POINT OF NO RETURN

Enter the following at the AutoCAD Command: prompt:

```
Command: dcircle
Specify center point for circles: ENTER (Pressing ENTER causes the program
    to do something unexpected.)
Application ERROR: Invalid type sent as command input
Application ERROR: Invalid type sent as command input
Command:
```

This is clearly a violation of the prime directive we stated earlier. No one could reasonably expect *Application ERROR: Invalid type sent as command input.*

 Note: It is not up to the programmer to decide if a user response is reasonable. Artful programs always behave reasonably, despite what the users do.

Why would users press ENTER? *Because they can.* What do users reasonably expect when they press ENTER? Let's see what happens with intrinsic AutoCAD commands.

Enter the following at the AutoCAD Command: prompt:

```
Command: circle   (Draws a 2-unit circle.)
Specify center point for circle or [3P/2P/Ttr (tan tan radius)]:
    ENTER
Point or option keyword required.
Specify center point for circle or [3P/2P/Ttr (tan tan radius)]:
    (Specify center point.)
Specify radius of circle or [Diameter] <default>: 2
```

In the case of the CIRCLE command, AutoCAD repeats the prompt after issuing an error message. I find it more reasonable to just to exit the command quietly.

Return to Visual LISP and edit (or load) the program shown in Listing 2–7.

Listing 2–7

```
;;; Filename: AP-02-07.lsp
(defun c:dcircle (/ p0 AP:SYSVARS AP:UNDOCTL AP:OLD-ERROR)
  (setvar "cmdecho" 0)                        ; quiet things down
```

```
(ap-push-error ap-error)                      ; save error handler
(ap-exit-transparent)                         ; exit if transparently
(ap-push-undo nil)                            ; start undo-group
(ap-push-vars '(("circlerad")))               ; save system variables
(setq p0                                      ; center of circles
     (getpoint "\nSpecify center point for circles: ")
)
(if (null p0)                                 ; exit if ENTER
  (exit)
)
(vl-cmdf "_circle" p0 1.0)                    ; draw the first circle
(vl-cmdf "_circle" p0 0.9)                    ; draw the second
(ap-pop-vars)                                 ; restore system variables
(ap-pop-undo)                                 ; end undo-group
(ap-pop-error)                                ; restore error handler
(princ)                                       ; exit quietly
)
```

Here, we'll avoid passing a null point to the **command** function by simply exiting the program.

 Note: As a rule, every user input must be checked for validity.

Select Tools|Load Text in Editor

Switch back to AutoCAD and enter the following at the Command: prompt:

```
Command: dcircle
Specify center point for circles: ENTER
Command:
```

Isn't that friendlier? We've gone about as far as we can go in crafting this little program.

DEM BONES, PART I

Before we go on, consider Listing 2–8.

Listing 2–8

```
;;; Filename: AP-02-08.lsp
(defun c:<command-name> (/ AP:SYSVARS AP:UNDOCTL AP:OLD-ERROR ...)
  (setvar "cmdecho" 0)                   ; quiet things down
  (ap-push-error ap-error)               ; save error handler
  (ap-exit-transparent)                  ; exit if transparently
  (ap-push-undo nil)                     ; start undo-group
  (ap-push-vars <var-list>)              ; save system variables
  (<command-name>-main)                  ; main function
  (ap-pop-vars)                          ; restore system variables
  (ap-pop-undo)                          ; end undo-group
  (ap-pop-error)                         ; restore error handler
  (princ)                                ; exit quietly
)
```

I started with the code shown in Listing 2–7 and generalized the code to produce the code skeleton shown above in Listing 2–8.

As you may infer, *command-name* is the name of the command we are defining. *var-list* is the list of system variables to be saved, and *command-name*-main is the main program.

 Tip: If you limit the size of each of your functions to that which you can see all at once in your editor (roughly 25 lines), the number of errors in your programs will diminish drastically.

We will be using this code skeleton (with subsequent improvements) in the programs we develop. What does this skeleton do for us?

- Eliminates unnecessary command line echoing.
- Saves and restores the AutoLISP error handler.
- Honors U and UNDO regardless of the system settings.
- Prevents commands from being run transparently.
- Honors the ESC key.
- Prevents AutoLISP from doing anything the user does not reasonably expect.

This is not too shabby, but we'll need something a little more complicated to expose some unexpected behavior in AutoLISP.

TAKING COMMAND OF COMMAND

Inside Visual LISP, create (or load) the program shown in Listing 2–9.

Listing 2–9

```
;;; Filename: AP-02-09.lsp
(defun c:cross (/ AP:SYSVARS AP:UNDOCTL AP:OLD-ERROR)
  (setvar "cmdecho" 0)                        ; quiet things down
  (ap-push-error ap-error)                    ; save error handler
  (ap-exit-transparent)                       ; exit if transparently
  (ap-push-undo nil)                          ; start undo-group
  (ap-push-vars nil)                          ; save system variables
  (cross-main)                                ; evoke main program
  (ap-pop-vars)                               ; restore system variables
  (ap-pop-undo)                               ; end undo-group
  (ap-pop-error)                              ; restore error handler
  (princ)                                     ; exit quietly
)
(defun cross-main (/ p0 p1 p2 p3 p4 size)
  (if (null                                   ; get center point
      (setq p0 (getpoint "\nSpecify center point for cross: "))
    )
   (exit                                      ; if no center point
  )
  (setq size 0.5)                             ; size of cross
  (setq p1 (ap-vector-dif p0 (list size 0 0)) ; left side
        p2 (ap-vector-sum p0 (list size 0 0)) ; right side
        p3 (ap-vector-dif p0 (list 0 size 0)) ; bottom
        p4 (ap-vector-sum p0 (list 0 size 0)) ; top
  )
  (vl-cmdf "_line" p1 p2 "")                   ; draw the first line
  (vl-cmdf "_line" p3 p4 "")                   ; draw the second
)
```

The command definition function, **c:cross,** is based on the skeleton shown in Listing 2–8. The **cross-main** function need not deal with the doing and undoing of doing business with AutoCAD.

The **cross-main** function uses the **ap-vector-sum** and **ap-vector-dif** functions, from the Artful Programming API, to compute the vector sums and differences, respectively, to locate the corners of a one-unit radius cross at the point specified by the user.

As we did for the **dcircle** command, we quietly exit if no center point is specified by the user.

Let's give it a try, and see if there's anything wrong with it.

Select Tools|Load Text in Editor.

Switch to AutoCAD, and enter the following at the Command: prompt:

```
Command: blipmode
Enter mode [ON/OFF] <default>: on (Sets BLIPMODE on.)
Command: cross
Specify center point for cross: 5,5 (Draws a cross at (5,5).)
```

Do you notice anything wrong here? If you look closely, you'll discover that AutoCAD has drawn blips at the ends of the lines it's drawn. The users expect to find blips only at the points that have been specified—to wit, the center of the cross (5,5). Let's dig a little deeper and see if there's anything else wrong here.

Enter the following at the AutoCAD Command: prompt:

```
Command: lastpoint (Examine LASTPOINT.)
Enter new value for LASTPOINT <5.0000,5.5000,0.0000>: ENTER
Command:
```

This point, (5,5.5,0), is not what we expected. AutoCAD users expect LASTPOINT to hold the last point they specified, not the last point passed to AutoCAD with the **command** or **vl-cmdf** function.

So far, we've found two subtle problems with our program. By now, you're asking yourself, "Is that all there is?" Unfortunately, no. (But then, you suspected as much, didn't you?)

Enter the following at the AutoCAD Command: prompt:

```
Command: -osnap
Current osnap modes: (Current modes are shown.)
Enter list of object snap modes: mid (Set a running object snap to
    midpoint.)
Command: osnapcoord
Enter new value for OSNAPCOORD <default>: 2 (Set keyboard priority for
    running object snaps.)
Command: line
Specify first point: 4,4
Specify next point or [Undo]: 4,6
Specify next point or [Undo]: (Draw a line from 4,4 to 4,6.)
Command: cross
Specify center point for cross: 4.5,6 (Applies object snap to left arm
    of cross.)
```

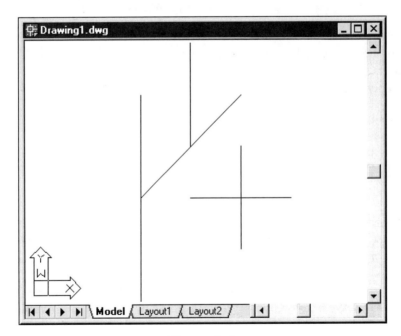

Figure 2–1

You'll get something like that shown in Figure 2–1. This certainly does not look right. What has happened is that AutoCAD has applied the running object snaps to the points passed to the **vl-cmdf** function.

Thus, we see that the **command** and **vl-cmdf** functions in and of themselves present a number of challenges to programmers who wish to create their own commands that work as the users expect them to work. These are summarized in Table 2–1.

Table 2–1

What AutoLISP Does	What Users Expect
If BLIPMODE is 1, AutoCAD will draw blips at every point passed to the **command** and vl-cmdf functions. This is annoying.	If BLIPMODE is 1, AutoCAD draws blips at points input by users.
If HIGHLIGHT is 1, objects selected by the **command** and **vl-cmdf** functions are highlighted. This is annoying, and slows things down.	If HIGHLIGHT is 1, only objects selected by the user are highlighted.
If OSNAPCOORD is set to a value other than 1, points passed to the **command** and **vl-cmdf** functions have object snaps applied to them. This is wrong.	Object snap settings have no effects other than on user picks.
Points passed to the **command** and **vl-cmdf** functions are saved in LASTPOINT. This is wrong.	Points input by users become LASTPOINT.

Before calls to the command function, we should

- Save the current values of BLIPMODE, HIGHLIGHT, OSNAPCOORD, and LASTPOINT.
- Set BLIPMODE to 0.
- Set HIGHLIGHT to 0.
- Set OSNAPCOORD to 1.

After calls to the command function, we should:

- **Restore the** prior values of BLIPMODE, HIGHLIGHT, OSNAPCOORD, and LASTPOINT.

This is easily accomplished with the **ap-push-command** and **ap-pop-command** functions from the Artful Programming API. We just use them to bracket calls to the **command** and **vl-cmdf** functions.

Return to Visual LISP, and edit (or load) the program shown in Listing 2–10.

Listing 2–10

```
;;; Filename: AP-02-10.lsp
(defun c:cross (/ AP:SYSVARS AP:UNDOCTL AP:OLD-ERROR)
  (setvar "cmdecho" 0)                     ; quiet things down
  (ap-push-error ap-error)                 ; save error handler
  (ap-exit-transparent)                    ; exit if transparently
  (ap-push-undo nil)                       ; start undo-group
  (ap-push-vars nil)                       ; save system variables
  (cross-main)                             ; evoke main program
  (ap-pop-vars)                            ; restore system variables
  (ap-pop-undo)                            ; end undo-group
  (ap-pop-error)                           ; restore error handler
  (princ)                                  ; exit quietly
)
(defun cross-main (/ p0 p1 p2 p3 p4 size)
  (if (null                                ; get center point
      (setq p0 (getpoint "\nSpecify center point for cross: "))
    )
    (exit)                                 ; if no center point
  )
  (setq size 0.5)                          ; size of cross
  (setq p1 (ap-vector-dif p0 (list size 0 0))  ; left side
        p2 (ap-vector-sum p0 (list size 0 0))  ; right side
        p3 (ap-vector-dif p0 (list 0 size 0))  ; bottom
        p4 (ap-vector-sum p0 (list 0 size 0))  ; top
  )
  (ap-push-command)                        ; save (command) vars
  (vl-cmdf "_line" p1 p2 "")               ; draw the first line
  (vl-cmdf "_line" p3 p4 "")               ; draw the second
  (ap-pop-command)                         ; restore (command) vars
)
```

Select Tools|Load Text in Editor.

Switch to AutoCAD, and enter the following at the Command: prompt:

```
Command: erase
Select objects: all
nnn found
Select objects: ENTER  (Clear the screen.)
Command: line
Specify first point: 4,4
Specify next point or [Undo]: 4,6
Specify next point or [Undo]: ENTER  (Draw a line from 4,4 to 4,6.)
Command: cross
Specify center point for cross: 4.5,6  (Draw the cross.)
```

Well, that looks better, but there's just one more problem with our little program.

Enter the following at the Command: prompt:

```
Command: lastpoint  (Examine LASTPOINT.)
Enter new value for LASTPOINT <4.5000,6.5000,0.0000>: ENTER
```

You would expect LASTPOINT to be (4.5,6,0), the last point that *you* specified. Apparently, the **getpoint** function does not set the LASTPOINT system variable. The **ap-getpoint** function, from the Artful Programming API, will take care of that.

Return to Visual LISP and edit (or load) the program shown in Listing 2–11.

Listing 2–11

```
;;; Filename: AP-02-11.lsp
(defun c:cross (/ AP:SYSVARS AP:UNDOCTL AP:OLD-ERROR)
  (setvar "cmdecho" 0)                  ; quiet things down
  (ap-push-error ap-error)              ; save error handler
  (ap-exit-transparent)                 ; exit if transparently
  (ap-push-undo nil)                    ; start undo-group
  (ap-push-vars nil)                    ; save system variables
  (cross-main)                          ; evoke main program
  (ap-pop-vars)                         ; restore system variables
  (ap-pop-undo)                         ; end undo-group
  (ap-pop-error)                        ; restore error handler
  (princ)                               ; exit quietly
```

```
)
(defun cross-main (/ p0 p1 p2 p3 p4 size)
  (if
    (null                                    ; get center point
      (setq p0 (ap-getpoint nil "\nSpecify center point for cross: "))
    )
    (exit)                                   ; if no center point
  )
  (setq size 0.5)                            ; size of cross
  (setq p1 (ap-vector-dif p0 (list size 0 0)) ; left side
        p2 (ap-vector-sum p0 (list size 0 0)) ; right side
        p3 (ap-vector-dif p0 (list 0 size 0)) ; bottom
        p4 (ap-vector-sum p0 (list 0 size 0)) ; top
  )
  (ap-push-command)                          ; save (command) vars
  (vl-cmdf "_line" p1 p2 "")                 ; draw the first line
  (vl-cmdf "_line" p3 p4 "")                 ; draw the second
  (ap-pop-command)                           ; restore (command) vars
)
```

Let's give it a try. Switch to AutoCAD and enter the following at the `Command:` prompt:

```
Command: erase
Select objects: all
nnn found
Select objects: ENTER (Clear the screen.)
Command: line
Specify first point: 4,4
Specify next point or [Undo]: 4,6
Specify next point or [Undo]: ENTER (Draw a line from 4,4 to 4,6)
Command: cross
Specify center point for cross: 4.5,6 (Draw the cross.)
Command: lastpoint (Examine LASTPOINT.)
Enter new value for LASTPOINT <4.5000,6.0000,0.0000>: ENTER
```

Absolutely wonderful! Here's what we've accomplished in this section:

- Eliminated extraneous blips.
- Eliminated extraneous highlighting.

- Compensated for running objects snaps.

- Updated the LASTPOINT system variable only when the user specifies a point with the **ap-getpoint** function.

Pat yourself on the back. Your programs are looking increasingly professional, but we still have more work to do.

NOTHING SELECTED, NOTHING GAINED

Inside Visual LISP, create (or load) the program shown in Listing 2–12.

Listing 2–12

```
;;; Filename: AP-02-12.lsp
(defun c:rot-10 (/ AP:SYSVARS AP:UNDOCTL AP:OLD-ERROR)
  (setvar "cmdecho" 0)                    ; quiet things down
  (ap-push-error ap-error)                ; save error handler
  (ap-exit-transparent)                   ; exit if transparently
  (ap-push-undo nil)                      ; start undo-group
  (ap-push-vars nil)                      ; save system variables
  (rot-10-main)                           ; evoke main program
  (ap-pop-vars)                           ; restore system variables
  (ap-pop-undo)                           ; end undo-group
  (ap-pop-error)                          ; restore error handler
  (princ)                                 ; exit quietly
)
(defun rot-10-main (/ n e-name ent p0 ss)
  (setq ss (ssget))                       ; select objects
  (ap-push-command)                       ; save (command) vars
  (setq n (sslength ss))                  ; number of objects
  (while (> n 0)
    (setq n (1- n))                       ; decrement counter
    (setq e-name (ssname ss n))           ; get entity name
    (setq ent (entget e-name))            ; get entity
    (setq p0 (ap-item 10 ent))            ; insertion point
    (vl-cmdf "_rotate" e-name "" p0 10)   ; rotate object
  )
  (ap-pop-command)                        ; restore (command) vars
)
```

This little program is supposed to rotate each selected text object 10° about its insertion point. We've incorporated all the artful tools we've learned so far. Note that the **ap-push-command** function is invoked after we've gotten our user inputs. The **ap-pop-command** function isn't evoked until all the **command** and **vl-cmdf** calls are completed. Let's give it a try.

Select Tools|Load Text in Editor.

Switch to AutoCAD. Draw yourself some single-line text entities and enter the following at the AutoCAD Command: prompt:

```
Command: rot-10
Select objects: (Select the objects.)
Select objects: (Rotates selected text entities.)
Command: rot-10
Select objects: ENTER
; error: bad argument type: lselsetp nil (Produces an error message.)
```

The error message is caused by **sslength** rejecting an empty selection set. This is easily fixed. The artful version of the **sslength** function, **ap-sslength**, returns zero if called with a null input.

Return to Visual LISP and edit (or load) the program shown in Listing 2–13.

Listing 2–13

```
;;; Filename: AP-02-13.lsp
(defun c:rot-10 (/ AP:SYSVARS AP:UNDOCTL AP:OLD-ERROR)
  (setvar "cmdecho" 0)                     ; quiet things down
  (ap-push-error ap-error)                 ; save error handler
  (ap-exit-transparent)                    ; exit if transparently
  (ap-push-undo nil)                       ; start undo-group
  (ap-push-vars nil)                       ; save system variables
  (rot-10-main)                            ; evoke main program
  (ap-pop-vars)                            ; restore system variables
  (ap-pop-undo)                            ; end undo-group
  (ap-pop-error)                           ; restore error handler
  (princ)                                  ; exit quietly
)
(defun rot-10-main (/ n e-name ent p0 ss)
  (setq ss (ssget))                        ; select objects
  (ap-push-command)                        ; save (command) vars
```

```
   (setq n (ap-sslength ss))              ; number of objects
   (while (> n 0)
     (setq n (1- n))                      ; decrement counter
     (setq e-name (ssname ss n))          ; get entity name
     (setq ent (entget e-name))           ; get entity
     (setq p0 (ap-item 10 ent))           ; insertion point
     (vl-cmdf "_rotate" e-name "" p0 10)  ; rotate object
   )
   (ap-pop-command)                       ; restore (command) vars
 )
```

Select Tools|Load Text in Editor.

Switch to AutoCAD and enter the following at the Command: prompt:

```
Command: rot-10
Select objects: ENTER
Command: (Terminates properly when nothing selected.)
```

That was easy.

 Tip: You must always prepare yourself for anything the user may do in response to a request for input.

PICK FIRST AND ASK QUESTIONS LATER

Your users may reasonably expect that your programs honor noun–verb object selection; you can pick things first, then call up the command. Let's give it a try.

Enter the following at the AutoCAD Command: prompt:

```
Command: pickfirst
Enter new value for PICKFIRST <default>: 1 (Ensure PICKFIRST is enabled.)
```

Select some text objects, then enter the following:

```
Command: rot-10  (Notice that as soon as you press ENTER, your selected objects are
   deselected.)
Select objects: ENTER
Command: (Ends ROT-10 command.)
```

Can we fix this? You bet!

Return to Visual LISP and edit (or load) the program shown in Listing 2–14.

Listing 2–14

```lisp
;;; Filename: AP-02-14.lsp
(defun c:rot-10 (/ ss AP:SYSVARS AP:UNDOCTL AP:OLD-ERROR)
  (setvar "cmdecho" 0)                        ; quiet things down
  (setq ss (ssget "I"))                       ; note implied selection set
  (ap-push-error ap-error)                    ; save error handler
  (ap-exit-transparent)                       ; exit if transparently
  (ap-push-undo nil)                          ; start undo-group
  (ap-push-vars nil)                          ; save system variables
  (rot-10-main ss)                            ; evoke main program
  (ap-pop-vars)                               ; restore system variables
  (ap-pop-undo)                               ; end undo-group
  (ap-pop-error)                              ; restore error handler
  (princ)                                     ; exit quietly
)
(defun rot-10-main (ss / n e-name ent p0)
  (if ss                                      ; use implied selection set
    (princ (strcat (itoa (sslength ss)) " found\n"))
    (setq ss (ssget))                         ; select objects
  )
  (ap-push-command)                           ; save (command) vars
  (setq n (ap-sslength ss))                   ; number of objects
  (while (> n 0)
    (setq n (1- n))                           ; decrement counter
    (setq e-name (ssname ss n))               ; get entity name
    (setq ent (entget e-name))                ; get entity
    (setq p0 (ap-item 10 ent))                ; insertion point
    (vl-cmdf "_rotate" e-name "" p0 10)       ; rotate object
  )
  (ap-pop-command)                            ; restore (command) vars
)
```

Select Tools|Load Text in Editor.

Switch back to AutoCAD, and while still at the Command: prompt, select four text objects, then enter the following:

```
Command: rot-10
4 found
Command:  (As soon as you press ENTER, the selected objects are rotated, and we're
          returned to the Command: prompt.)
```

There, we've implemented noun–verb object selection.

Now let's examine what happens when you try to access the previous selection set.

YOU MUST HAVE BEEN GOOD IN A PREVIOUS SELECTION SET

While still in AutoCAD, enter the following at the Command: prompt:

```
Command: rot-10
Select objects: p
1 found
Select objects: ENTER
Command:  (AutoCAD finds only one object in the previous selection set.)
```

What has happened here is that the **command** and **vl-cmdf** functions create a new previous selection set *each time* they execute an AutoCAD command that creates one. In this specific instance, it's the **(vl-cmdf "_rotate")** function call.

Thus, it is up to the artful programmer to *explicitly* create a previous selection set after all the calls to the command function are made. This should occur before the final call to **ap-pop-command**. Let's give it a try.

Return to Visual LISP and edit (or load) the program shown in Listing 2–15.

Listing 2–15

```
;;; Filename: AP-02-15.lsp
(defun c:rot-10 (/ ss AP:SYSVARS AP:UNDOCTL AP:OLD-ERROR)
  (setvar "cmdecho" 0)                    ; quiet things down
  (setq ss (ssget "I"))                   ; note implied selection set
  (ap-push-error ap-error)                ; save error handler
  (ap-exit-transparent)                   ; exit if transparently
  (ap-push-undo nil)                      ; start undo-group
  (ap-push-vars nil)                      ; save system variables
  (rot-10-main ss)                        ; evoke main program
  (ap-pop-vars)                           ; restore system variables
```

```
  (ap-pop-undo)                            ; end undo-group
  (ap-pop-error)                           ; restore error handler
  (princ)                                  ; exit quietly
)
(defun rot-10-main (ss / n e-name ent p0)
  (if ss                                   ; use implied selection set
    (princ (strcat (itoa (sslength ss)) " found\n"))
    (setq ss (ssget))                      ; select objects
  )
  (ap-push-command)                        ; save (command) vars
  (setq n (ap-sslength ss))                ; number of objects
  (while (> n 0)
    (setq n (1- n))                        ; decrement counter
    (setq e-name (ssname ss n))            ; get entity name
    (setq ent (entget e-name))             ; get entity
    (setq p0 (ap-item 10 ent))             ; insertion point
    (vl-cmdf "_rotate" e-name "" p0 10)    ; rotate object
  )
  (ap-ssprevious ss)                       ; create Previous sset
  (ap-pop-command)                         ; restore (command) vars
)
```

The call to the **ap-ssprevious** function from the Artful Programming API establishes *ss* as the previous selection set.

Select Tools|Load Text in Editor.

Switch to AutoCAD, and enter the following at the Command: prompt:

```
Command: rot-10
Select objects: (Pick four text objects.)
4 found
Select objects: ENTER (Rotates four text objects.)
Command: rot-10
Select objects: p
4 found
Select objects: ENTER
Command: (Treats four text objects previously selected as the previous selection set.)
```

That fixed that! Is there anything *else* wrong with this little program? By now, you've guessed the answer: Yes, of course there is.

RULING THE WORLD COORDINATE SYSTEM

Let's see how our program deals with a User Coordinate System other than the World UCS.

Enter the following at the AutoCAD Command: prompt:

```
Command: ucs
Current ucs name:  *WORLD* (Move the UCS origin.)
Enter an option [New/Move/orthoGraphic/Prev/Restore/Save/Del/
   Apply/?/World] <World>: m
Specify new origin point or [Zdepth]<0,0,0>: 10,0,0
Command: rot-10
Select objects: (Select the objects.)
Select objects: ENTER
Command: (Objects are rotated, but not quite right.)
```

The objects you selected were not rotated about their insertion points. That's because the insertion points stored with each object are in the World Coordinate System (WCS), whereas the **(command)** function works in the current User Coordinate System (UCS).

Because each object is to be rotated about its insertion point, we'll just set the UCS to each object's UCS before rotating the object and put it back after rotating it.

Return to Visual LISP and edit (or load) the program shown in Listing 2–16.

Listing 2–16

```
;;; Filename: AP-02-16.lsp
(defun c:rot-10 (/ ss AP:SYSVARS AP:UNDOCTL AP:OLD-ERROR)
  (setvar "cmdecho" 0)                     ; quiet things down
  (setq ss (ssget "I"))                    ; note implied selection set
  (ap-push-error ap-error)                 ; save error handler
  (ap-exit-transparent)                    ; exit if transparently
  (ap-push-undo nil)                       ; start undo-group
  (ap-push-vars nil)                       ; save system variables
  (rot-10-main ss)                         ; evoke main program
  (ap-pop-vars)                            ; restore system variables
  (ap-pop-undo)                            ; end undo-group
  (ap-pop-error)                           ; restore error handler
  (princ)                                  ; exit quietly
)
```

```
(defun rot-10-main (ss / n e-name)
  (if ss                                    ; use implied selection set
    (princ (strcat (itoa (sslength ss)) " found\n"))
    (setq ss (ssget))                       ; select objects
  )
  (ap-push-command)                         ; save (command) vars
  (setq n (ap-sslength ss))                 ; number of objects
  (while (> n 0)
    (setq n (1- n))                         ; decrement counter
    (setq e-name (ssname ss n))             ; get entity name
    (vl-cmdf "_ucs" "_n" "_ob" e-name)      ; set ucs to that of object
    (vl-cmdf "_rotate" e-name "" '(0 0 0) 10) ; rotate object
    (vl-cmdf "_ucs" "_p")                   ; restore ucs
  )
  (ap-ssprevious ss)                        ; create Previous sset
  (ap-pop-command)                          ; restore (command) vars
)
```

Select Tools|Load Text in Editor.

Switch to AutoCAD. You should still be in the UCS that you defined last time. Enter the following at the Command: prompt:

```
Command: rot-10
Select objects: (Select the objects.)
Select objects: ENTER (Rotates objects properly.)
Command: ucs
Current ucs name:  *NO NAME* (Return to the World UCS.)
Enter an option [New/Move/orthoGraphic/Prev/Restore/Save/Del/
  Apply/?/World] <World>: ENTER
Command:
```

Now each object is rotated about its insertion point, but we've introduced some more unexpected behavior.

First, the UCS icon bounces all over the place as it moves from one object to another. Users don't expect to see this, so let's fix it. We'll just turn off the UCS icon display before we start bouncing around.

Return to Visual LISP and edit (or load) the program shown in Listing 2–17.

Listing 2–17

```
;;; Filename: AP-02-17.lsp
(defun c:rot-10 (/ ss AP:SYSVARS AP:UNDOCTL AP:OLD-ERROR)
  (setvar "cmdecho" 0)                      ; quiet things down
  (setq ss (ssget "I"))                     ; note implied selection set
  (ap-push-error ap-error)                  ; save error handler
  (ap-exit-transparent)                     ; exit if transparently
  (ap-push-undo nil)                        ; start undo-group
  (ap-push-vars nil)                        ; save system variables
  (rot-10-main ss)                          ; evoke main program
  (ap-pop-vars)                             ; restore system variables
  (ap-pop-undo)                             ; end undo-group
  (ap-pop-error)                            ; restore error handler
  (princ)                                   ; exit quietly
)
(defun rot-10-main (ss / n e-name)
  (if ss                                    ; use implied selection set
    (princ (strcat (itoa (sslength ss)) " found\n"))
    (setq ss (ssget))                       ; select objects
  )
  (ap-push-vars '(("ucsicon" . 0)))         ; hide ucsicon
  (ap-push-command)                         ; save (command) vars
  (setq n (ap-sslength ss))                 ; number of objects
  (while (> n 0)
    (setq n (1- n))                         ; decrement counter
    (setq e-name (ssname ss n))             ; get entity name
    (vl-cmdf "_ucs" "_n" "_ob" e-name)      ; set ucs to that of object
    (vl-cmdf "_rotate" e-name "" '(0 0 0) 10); rotate object
    (vl-cmdf "_ucs" "_p")                    ; restore ucs
  )
  (ap-ssprevious ss)                        ; create Previous sset
  (ap-pop-command)                          ; restore (command) vars
)
```

Select Tools|Load Text in Editor.

Return to AutoCAD, create a 20 by 20 array of text, and try the **rot-10** command
again. This time, we've turned off the ucs icon for the duration of the command.
It's a lot less distracting and saves a little time. We have just one more problem
with our command.

Enter the following at the AutoCAD Command: prompt:

```
Command: 3dpoly
Specify start point of polyline: 0,0
Specify endpoint of line or [Undo]: 4,0
Specify endpoint of line or [Undo]: 4,3
Specify endpoint of line or [Close/Undo]: c (Draw a 3D polyline.)
Command: rot-10
Select objects: Select 3D polyline.
Select objects: ENTER
3D Polyline object does not define a coordinate system
Command:
```

What happened?

As soon as we're done selecting objects, we got an error when we tried setting our UCS to that of the 3D polyline (which doesn't have one, according to the error message).

You see, although our command was designed to rotate text objects, it makes no effort to validate that the user didn't select something other than text.

 Tip: You can never guarantee that a user will pick what you ask them to pick.

The fix is fairly easy, though somewhat obscure. Although our command is designed to rotate text objects, it makes no effort to validate that the user didn't select something other than text.

THE OBJECTS OF OUR DESIRES

We're going to use selection set filtering to direct the **ssget** function to reject (filter out) anything the user selects other than text entities. Let's do it.

Return to Visual LISP and edit (or load) the program shown in Listing 2–18.

Listing 2–18

```
;;; Filename: AP-02-18.lsp
(defun c:rot-10 (/ ss FILTER AP:SYSVARS AP:UNDOCTL AP:OLD-ERROR)
  (setvar "cmdecho" 0)                      ; quiet things down
  (setq FILTER '((0 . "TEXT")))             ; text filter
```

```
    (setq ss (ssget "I" FILTER))           ; select only text
    (ap-push-error ap-error)               ; save error handler
    (ap-exit-transparent)                  ; exit if transparently
    (ap-push-undo nil)                     ; start undo-group
    (ap-push-vars nil)                     ; save system variables
    (rot-10-main ss)                       ; evoke main program
    (ap-pop-vars)                          ; restore system variables
    (ap-pop-undo)                          ; end undo-group
    (ap-pop-error)                         ; restore error handler
    (princ)                                ; exit quietly
)
(defun rot-10-main (ss / n e-name)
  (if ss                                   ; use implied selection set
    (princ (strcat (itoa (sslength ss)) " found\n"))
    (setq ss (ssget FILTER))               ; select only text
  )
  (ap-push-vars '(("ucsicon" . 0)))        ; hide ucsicon
  (ap-push-command)                        ; save (command) vars
  (setq n (ap-sslength ss))                ; number of objects
  (while (> n 0)
    (setq n (1- n))                        ; decrement counter
    (setq e-name (ssname ss n))            ; get entity name
    (vl-cmdf "_ucs" "_n" "_ob" e-name)     ; set ucs to that of object
    (vl-cmdf "_rotate" e-name "" '(0 0 0) 10); rotate object
    (vl-cmdf "_ucs" "_p")                  ; restore ucs
  )
  (ap-ssprevious ss)                       ; create Previous sset
  (ap-pop-command)                         ; restore (command) vars
)
```

Select Tools|Load Text in Editor.

Return to AutoCAD and enter the following at the Command: prompt:

```
Command: rot-10 (Now the program filters out everything selected other than text.)
Select objects: (Select 3D polyline.)
1 found, 1 was filtered out.
Select objects: (Select text.)
1 found, 1 total
Select objects: ENTER
Command:
```

Now, any selections of any entity other that text will be filtered out, and your program no longer needs to deal with anything unexpected.

Here's what we've accomplished in this section:

- Eliminated program errors when the user selects nothing.
- Honored noun–verb object selection.
- Created a previous selection set.
- Eliminated program errors when the user selects undesired objects.

Once more, pat yourself on the back. Your programs are looking even more professional, but we still have more to do.

VALIDATING AND RESTRICTING INPUTS

Let's start out by examining a little program with the job of determining if three distances supplied by the user represent an equilateral (three sides equal), isosceles (precisely two sides equal), or scalene (three sides different) triangle.

Inside Visual LISP, create (or load) the program shown in Listing 2–19.

Listing 2–19

```
;;; Filename: AP-02-19.lsp
(defun c:tri-me (/ AP:SYSVARS AP:UNDOCTL AP:OLD-ERROR)
  (setvar "cmdecho" 0)                    ; quiet things down
  (ap-push-error ap-error)                ; save error handler
  (ap-exit-transparent)                   ; exit if transparently
  (ap-push-undo nil)                      ; start undo-group
  (ap-push-vars nil)                      ; save system variables
  (tri-me-main)                           ; evoke main program
  (ap-pop-vars)                           ; restore system variables
  (ap-pop-undo)                           ; end undo-group
  (ap-pop-error)                          ; restore error handler
  (princ)                                 ; exit quietly
)
(defun tri-me-main (/ d1 d2 d3)
  (setq d1 (getdist "\nSpecify length of first side:  "))
  (setq d2 (getdist "\nSpecify length of second side:  "))
  (setq d3 (getdist "\nSpecify length of third side:  "))
  (cond ((= d1 d2 d3)                        ; equilateral
         (princ "\nThis is an Equilateral triangle.")
         )
        ((or (= d1 d2) (= d2 d3) (= d3 d1)) ; isosceles
         (princ "\nThis is an Isosceles triangle.")
         )
```

```
        (t                                    ; scalene
      (princ "\nThis is a Scalene triangle.")
      )
  )
)
```

The program looks like it should work—we've incorporated all the artful tools we've learned so far. Let's try it out.

Select Tools|Load Text in Editor.

Return to AutoCAD, and enter the following at the Command: prompt:

```
Command: tri-me Specify length of first side:  2
Specify length of second side: 2
Specify length of third side:  2
This is an Equilateral triangle.
```

Our program has successfully recognized an equilateral triangle. Let's see how it does with isosceles. We'll have to test three possible cases.

```
Command: tri-me Specify length of first side:  2
Specify length of second side: 2
Specify length of third side:  1
This is an Isosceles triangle.
Command: tri-me Specify length of first side:  2
Specify length of second side: 1
Specify length of third side:  2
This is an Isosceles triangle.
Command: tri-me Specify length of first side:  1
Specify length of second side: 2
Specify length of third side:  2
This is an Isosceles triangle.
```

We've recognized all three cases of isosceles triangles here. Let's try a scalene triangle now.

```
Command: tri-me Specify length of first side:  3
Specify length of second side: 4
Specify length of third side:  5
This is a Scalene triangle.
```

All of our test cases worked, but is there anything wrong with our program?

Try out the following.

```
Command: tri-me Specify length of first side: ENTER
Specify length of second side: ENTER
Specify length of third side: ENTER
This is an Equilateral triangle. (No, it isn't!)
```

If the user presses ENTER at the first prompt, they probably don't want to enter the sides of a triangle; they probably want to abort the command. Let's help them.

Return to Visual LISP and edit (or load) the program shown in Listing 2–20.

Listing 2–20

```
;;; Filename: AP-02-20.lsp
(defun c:tri-me (/ AP:SYSVARS AP:UNDOCTL AP:OLD-ERROR)
  (setvar "cmdecho" 0)                    ; quiet things down
  (ap-push-error ap-error)                ; save error handler
  (ap-exit-transparent)                   ; exit if transparently
  (ap-push-undo nil)                      ; start undo-group
  (ap-push-vars nil)                      ; save system variables
  (tri-me-main)                           ; evoke main program
  (ap-pop-vars)                           ; restore system variables
  (ap-pop-undo)                           ; end undo-group
  (ap-pop-error)                          ; restore error handler
  (princ)                                 ; exit quietly
)
(defun tri-me-main (/ d1 d2 d3)
  (setq d1 (getdist "\nSpecify length of first side: "))
  (if (not d1)                            ; user pressed ENTER
    (exit)
  )
  (setq d2 (getdist "\nSpecify length of second side: "))
  (setq d3 (getdist "\nSpecify length of third side: "))
  (cond ((= d1 d2 d3)                     ; equilateral
         (princ "\nThis is an Equilateral triangle.")
         )
        ((or (= d1 d2) (= d2 d3) (= d3 d1)) ; isosceles
         (princ "\nThis is an Isosceles triangle.")
         )
        (t                               ; scalene
```

```
        (princ "\nThis is a Scalene triangle.")
        )
    )
)
```

Let's give it a try.

Select Tools|Load Text in Editor.

Return to AutoCAD, and enter the following at the Command: prompt:

```
Command: tri-me
Specify length of first side: ENTER
Command: (We just return to the Command: prompt.)
```

That worked nicely, but are we done? Let's find out.

```
Command: tri-me
Specify length of first side: 2
Specify length of second side: ENTER
Specify length of third side: ENTER
This is an Isosceles triangle (This doesn't look like a triangle to me.)
```

It isn't a triangle. Let's fix this problem now.

Return to Visual LISP and edit (or load) the program shown in Listing 2–21.

Listing 2–21

```
;;; Filename: AP-02-21.lsp
(defun c:tri-me (/ AP:SYSVARS AP:UNDOCTL AP:OLD-ERROR)
  (setvar "cmdecho" 0)                    ; quiet things down
  (ap-push-error ap-error)                ; save error handler
  (ap-exit-transparent)                   ; exit if transparently
  (ap-push-undo nil)                      ; start undo-group
  (ap-push-vars nil)                      ; save system variables
  (tri-me-main)                           ; evoke main program
  (ap-pop-vars)                           ; restore system variables
  (ap-pop-undo)                           ; end undo-group
  (ap-pop-error)                          ; restore error handler
  (princ)                                 ; exit quietly
)
```

```
(defun tri-me-main (/ d1 d2 d3)
  (setq d1 (getdist "\nSpecify length of first side:  "))
  (if (not d1)                              ; user pressed ENTER
    (exit)
  )
  (initget 1)                               ; disallow null input
  (setq d2 (getdist "\nSpecify length of second side: "))
  (initget 1)                               ; disallow null input
  (setq d3 (getdist "\nSpecify length of third side:  "))
  (cond ((= d1 d2 d3)                       ; equilateral
      (princ "\nThis is an Equilateral triangle.")
      )
      ((or (= d1 d2) (= d2 d3) (= d3 d1)) ; isosceles
      (princ "\nThis is an Isosceles triangle.")
      )
      (t                                   ; scalene
      (princ "\nThis is a Scalene triangle.")
      )
  )
)
```

The calls to the **initget** function disallow null inputs in the subsequent calls to the **getdist** function. Let's try it out.

Select Tools|Load Text in Editor.

Return to AutoCAD, and enter the following at the Command: prompt:

```
Command: tri-me
Specify length of first side: 2
Specify length of second side: ENTER
Requires numeric distance or two points. (AutoCAD rejects the null input
    and tells you why.)
Specify length of second side: 2 (AutoCAD asks again.)
Specify length of third side: ENTER
Requires numeric distance or two points. (AutoCAD rejects the null input
    and tells you why.)
Specify length of third side: 2 (AutoCAD asks again...)
This is an Equilateral triangle  (...and gives the right answer.)
```

Are we done? Not yet. Let's see why.

```
Command: tri-me
Specify length of first side: 0
Specify length of second side: 0
Specify length of third side: 0
This is an Equilateral triangle. (No, it isn't.)
Command: tri-me
Specify length of first side: -1
Specify length of second side: -1
Specify length of third side: -1
This is an Equilateral triangle. (No it isn't.)
```

In both cases, although the sides are equal, we've not specified a triangle. We'll use the **initget** function to limit our inputs to nonzero, non-negative values.

Return to Visual LISP and edit (or load) the program shown in Listing 2–22.

Listing 2–22

```
;;; Filename: AP-02-22.lsp
(defun c:tri-me (/ AP:SYSVARS AP:UNDOCTL AP:OLD-ERROR)
  (setvar "cmdecho" 0)                          ; quiet things down
  (ap-push-error ap-error)                       ; save error handler
  (ap-exit-transparent)                          ; exit if transparently
  (ap-push-undo nil)                             ; start undo-group
  (ap-push-vars nil)                             ; save system variables
  (tri-me-main)                                  ; evoke main program
  (ap-pop-vars)                                  ; restore system variables
  (ap-pop-undo)                                  ; end undo-group
  (ap-pop-error)                                 ; restore error handler
  (princ)                                        ; exit quietly
)
(defun tri-me-main (/ d1 d2 d3)
  (initget (+ 2 4))                              ; no zero, negative
  (setq d1 (getdist "\nSpecify length of first side: "))
  (if (not d1)                                   ; user pressed ENTER
    (exit)
  )
  (initget (+ 1 2 4))                            ; no null, zero, negative
  (setq d2 (getdist "\nSpecify length of second side: "))
```

```
(initget (+ 1 2 4))                          ; no null, zero, negative
(setq d3 (getdist "\nSpecify length of third side: "))
(cond ((= d1 d2 d3)                          ; equilateral
      (princ "\nThis is an Equilateral triangle.")
      )
     ((or (= d1 d2) (= d2 d3) (= d3 d1)) ; isosceles
      (princ "\nThis is an Isosceles triangle.")
      )
     (t                                      ; scalene
      (princ "\nThis is a Scalene triangle.")
      )
    )
  )
)
```

Let's give it a try.

Select Tools|Load Text in Editor.

Return to AutoCAD, and enter the following at the Command: prompt:

Command: **tri-me**
Specify length of first side: **0**
Value must be positive and nonzero. *(You can't get past here with negative or zero.)*
Specify length of first side: **1**
Specify length of second side: ENTER *(You can't get past here with null, negative, or zero.)*
Requires numeric distance or two points.
Specify length of second side: **0**
Value must be positive and nonzero.
Specify length of second side: **1**
Specify length of third side: **-1**
Value must be positive and nonzero. *(You can't get past here with null, negative, or zero.)*
Specify length of third side: **1**
This is an Equilateral triangle.

Pretty spiffy, huh? Is that it? Well, there's one last case that isn't covered.

Command: **tri-me** *(Spot the bug?)*

```
Specify length of first side:  2
Specify length of second side: 2
Specify length of third side:  10
This is an Isosceles triangle.
```

Unfortunately, it isn't an isosceles triangle—it isn't a triangle at all. In order for three distances, d1, d2, and d3, to specify a triangle, the following relationship must hold:

$$|d1 - d2| < d3 < (d1 + d2)$$

I would be very surprised if this test were built into the **initget** function. In fact, I'd be a little worried.

When you have application-specific constraints, you're going to have to write application-specific functions. In other words, if you're going to do something unusual, you're going to have to do it yourself.

Return to Visual LISP, and edit (or load) the program shown in Listing 2–23.

Listing 2–23

```
;;; Filename: AP-02-23.lsp
(defun c:tri-me (/ AP:SYSVARS AP:UNDOCTL AP:OLD-ERROR)
  (setvar "cmdecho" 0)                    ; quiet things down
  (ap-push-error ap-error)                ; save error handler
  (ap-exit-transparent)                   ; exit if transparently
  (ap-push-undo nil)                      ; start undo-group
  (ap-push-vars nil)                      ; save system variables
  (tri-me-main)                           ; evoke main program
  (ap-pop-vars)                           ; restore system variables
  (ap-pop-undo)                           ; end undo-group
  (ap-pop-error)                          ; restore error handler
  (princ)                                 ; exit quietly
)
(defun tri-me-main (/ d1 d2 d3)
  (initget (+ 2 4))                       ; no zero, negative
  (setq d1 (getdist "\nSpecify length of first side: "))
  (if (not d1)                            ; user pressed ENTER
    (exit)
  )
  (initget (+ 1 2 4))                     ; no null, zero, negative
```

```
(setq d2 (getdist "\nSpecify length of second side: "))
(setq d3 (tri-me-getd3 d1 d2))              ; get third side
(cond ((= d1 d2 d3)                         ; equilateral
       (princ "\nThis is an Equilateral triangle.")
       )
      ((or (= d1 d2) (= d2 d3) (= d3 d1)) ; isosceles
       (princ "\nThis is an Isosceles triangle.")
       )
      (t                                    ; scalene
       (princ "\nThis is a Scalene triangle.")
       )
  )
)
(defun tri-me-getd3 (d1 d2 / d3 mind3 maxd3)
  (setq mind3 (abs (- d1 d2)))              ; minimum d3
  (setq maxd3 (+ d1 d2))                    ; maximum d3
  (while (progn (initget 1)                 ; no null
             (setq d3 (getdist "\nSpecify length of third side:  "))
             (not (< mind3 d3 maxd3))     ; too short or too long
         )
    (princ (strcat "\nRequires a distance greater than "
                   (rtos mind3)
                   " and less than "
                   (rtos maxd3)
                   ".\n"
            )
      )
    )
  )
  d3
)
```

We've relegated the getting of d3 to a separate function, **tri-me-getd3**. This helps keep our defuns short enough to be viewed all at once in our editor.

Select Tools|Load Text in Editor.

Return to AutoCAD, and enter the following at the Command: prompt:

```
Command: tri-me
Specify length of first side: 2
Specify length of second side: 2
```

```
Specify length of third side:  10
Requires a distance greater than 0.0000 and less than 4.0000.
Specify length of third side:  3
This is an Isosceles triangle.
```

And *that's* an isosceles triangle. The "0.0000" and "4.0000" don't look quite right. We'll use the DIMZIN dimensioning variable to suppress trailing zeros. Specifically, we want to set the 8 bit of DIMZIN without disturbing any other bits that are set. We'll be sure to put them back when we're done.

Return to Visual LISP and edit (or load) the program shown in Listing 2–24.

Listing 2–24

```
;;; Filename: AP-02-24.lsp
(defun c:tri-me (/ AP:SYSVARS AP:UNDOCTL AP:OLD-ERROR)
  (setvar "cmdecho" 0)                          ; quiet things down
  (ap-push-error ap-error)                      ; save error handler
  (ap-exit-transparent)                         ; exit if transparently
  (ap-push-undo nil)                            ; start undo-group
  (ap-push-vars                                 ; save system variables
    (list (cons "dimzin" (logior (getvar "dimzin") 8)))
                                                ; Suppress trailing zeros
  )
  (tri-me-main)                                 ; evoke main program
  (ap-pop-vars)                                 ; restore system variables
  (ap-pop-undo)                                 ; end undo-group
  (ap-pop-error)                                ; restore error handler
  (princ)                                       ; exit quietly
)
(defun tri-me-main (/ d1 d2 d3)
  (initget (+ 2 4))                             ; no zero, negative
  (setq d1 (getdist "\nSpecify length of first side: "))
  (if (not d1)                                  ; user pressed ENTER
    (exit)
  )
  (initget (+ 1 2 4))                           ; no null, zero, negative
  (setq d2 (getdist "\nSpecify length of second side: "))
  (setq d3 (tri-me-getd3 d1 d2))                ; get third side
  (cond ((= d1 d2 d3)                           ; equilateral
      (princ "\nThis is an Equilateral triangle.")
```

```
          )
          ((or (= d1 d2) (= d2 d3) (= d3 d1)) ; isosceles
          (princ "\nThis is an Isosceles triangle.")
          )
          (t                                   ; scalene
          (princ "\nThis is a Scalene triangle.")
          )
      )
  )
  (defun tri-me-getd3 (d1 d2 / d3 mind3 maxd3)
    (setq mind3 (abs (- d1 d2)))               ; minimum d3
    (setq maxd3 (+ d1 d2))                     ; maximum d3
    (while (progn (initget 1)                  ; no null
              (setq d3 (getdist "\nSpecify length of third side:  "))
              (not (< mind3 d3 maxd3))     ; too short or too long
          )
      (princ (strcat "\nRequires a distance greater than "
                  (rtos mind3)
                  " and less than "
                  (rtos maxd3)
                  ".\n"
          )
      )
    )
    d3
  )
```

Select Tools|Load Text in Editor.

Return to AutoCAD, and enter the following at the Command: prompt:

```
Command: tri-me
Specify length of first side:  2
Specify length of second side:  2
Specify length of third side:  10
Requires a distance greater than 0 and less than 4.
Specify length of third side:  3
This is an Isosceles triangle.
```

That looks better, but are we done? By now, you should have guessed that the answer is still no.

IT AIN'T DEFAULT OF AUTOCAD

Most of AutoCAD's intrinsic commands allow for default values; something you get when you press ENTER. Instead of having to reenter the same data, most AutoCAD commands assume you will be using the last value you used. Although this may not always be a valid assumption, there is at least a chance that you'll be using the same value you used last time.

AutoCAD abounds with examples of defaults. Because our goal is to act like AutoCAD, let's see what we're trying to emulate.

```
Command: polygon
Enter number of sides <4>: ENTER
Specify center of polygon or [Edge]: (Pick a point.)
Enter an option [Inscribed in circle/Circumscribed about circle]
   <I>: ENTER
Specify radius of circle: (Specify the radius.)
```

Specifically, the POLYGON command remembers the number of sides you entered last time, as well as whether the polygon was to be inscribed or circumscribed.

In our case, we *could* remember the values given for each of our distances. It certainly won't hurt the users' productivity if we do. Perhaps someone will someday even thank you for providing defaults. (If no one else does, *I* will.)

The question then arises of where to save our defaults. Given the docucentric nature of AutoCAD 2000, it makes sense to store them in the document (.dwg file) itself. Once you get used to defaults transcending AutoCAD sessions, you'll wonder how you ever got by without them.

Return to Visual LISP and edit (or load) the program shown in Listing 2–25.

Listing 2–25

```
;;; Filename: AP-02-25.lsp
(defun c:tri-me (/ AP:DICTNAME AP:SYSVARS AP:UNDOCTL AP:OLD-ERROR)
  (setq AP:DICTNAME "tri-me")                ; Dictionary Name
  (setvar "cmdecho" 0)                       ; quiet things down
  (ap-push-error ap-error)                   ; save error handler
  (ap-exit-transparent)                      ; exit if transparently
  (ap-push-undo nil)                         ; start undo-group
  (ap-push-vars                              ; save system variables
    (list (cons "dimzin" (logior (getvar "dimzin") 8)))) ; Suppress
```

```
    trailing zeros
    )
    (tri-me-main)                          ; evoke main program
    (ap-pop-vars)                          ; restore system variables
    (ap-pop-undo)                          ; end undo-group
    (ap-pop-error)                         ; restore error handler
    (princ)                                ; exit quietly
  )
(defun tri-me-main (/ d1 d2 d3)
  (mapcar 'ap-getdict '(d1 d2 d3))         ; get defaults
  (initget (+ 2 4))                        ; no zero, negative
  (setq d1 (ap-getdist nil "\nSpecify length of first side" d1))
  (if (not d1)                             ; user pressed ENTER
    (exit)
  )
  (initget (+ 2 4))                        ; no zero, negative
  (setq d2 (ap-getdist nil "\nSpecify length of second side" d2))
                                           ; get second side
  (setq d3 (tri-me-getd3 d1 d2 d3))        ; get third side
  (mapcar 'ap-putdict '(d1 d2 d3))         ; save defaults
  (cond ((= d1 d2 d3)                      ; equilateral
         (princ "\nThis is an Equilateral triangle.")
         )
        ((or (= d1 d2) (= d2 d3) (= d3 d1)) ; isosceles
         (princ "\nThis is an Isosceles triangle.")
         )
        (t                                 ; scalene
         (princ "\nThis is a Scalene triangle.")
         )
  )
)
(defun tri-me-getd3 (d1 d2 d3 / mind3 maxd3)
  (setq mind3 (abs (- d1 d2)))             ; minimum d3
  (setq maxd3 (+ d1 d2))                   ; maximum d3
  (while
    (progn (setq d3 (ap-getdist nil "\nSpecify length of third side" d3))
           (not (< mind3 d3 maxd3))        ; too short or too long
    )(princ (strcat "\nRequires a distance greater than "
                    (rtos mind3)
                    " and less than "
```

```
                    (rtos maxd3)
                    ".\n"
                )
            )
        )
        d3
    )
```

The global variable (for this program) `AP:DICTNAME` is the name of the drawing dictionary in which to save and restore our defaults. The **ap-getdict** function from the Artful Programming API restores the specified AutoLISP variable from the dictionary specified by `AP:DICTNAME`. The **ap-putdict** function, also from the Artful Programming API, saves the specified AutoLISP variable to the dictionary specified by `AP:DICTNAME`.

The **ap-getdist** function, again from the Artful Programming API, adds default support to the **getdist** function. In addition to the default-friendly **ap-getdist** function, the Artful Programming utilities provide the default-friendly **ap-getreal**, **ap-getint**, **ap-getkword**, and **ap-getstring** functions.

 Note: We restore our defaults at the start of the main function and don't save them until we've gotten all our inputs. By doing so, we won't affect our defaults if we cancel before completing the command.

Let's give it a try.

Select Tools|Load Text in Editor.

Return to AutoCAD, and enter the following at the `Command:` prompt:

```
Command: tri-me  (Nothing different so far.)
Specify length of first side: 3
Specify length of second side: 4
Specify length of third side: 5
This is a Scalene triangle.  (Nothing different so far.)
Command: tri-me  (Here we've saved two whole keystrokes.)
Specify length of first side <3>: ENTER
Specify length of second side <4>: ENTER
Specify length of third side <5>: 3
This is an Isosceles triangle.  (Here we've saved two whole keystrokes.)
```

The next time you run the **tri-me** command within this drawing, it will use the defaults of 3, 4, and 3, respectively. Amazingly enough, the defaults will be maintained even if you exit AutoCAD.

Once more, we've gone about as far as we can go with this program. What have we accomplished?

- Validated and restricted inputs to prevent crashes.
- Allowed for default values.
- Saved the default values with the drawing file.

KEYWORDS, PROMPTS, AND CONTEXT MENUS

If you've been wondering about the relationship between keywords, prompts, and context menus in AutoCAD 2000, it's easier than it looks.

Enter the following at AutoCAD's Command: prompt:

```
Command: shortcutmenu
Enter new value for SHORTCUTMENU <default>: 11 (Establish default
    action.)
Command: circle
Specify center point for circle or [3P/2P/Ttr (tan tan radius)]:
    (Right-click on graphics area. Displays shortcut menu.)
```

Right-clicking on the graphics area while command options are present pops up a shortcut menu, as shown in Figure 2–2.

Figure 2–2

Notice that the middle section of the menu contains those options that appear on the command line. When you're done admiring the shortcut menu, select Cancel.

How does AutoCAD differentiate between keywords and anything else that appears on the command line? The seems to be just three rules:

- All keywords must appear between square brackets ([and]).

- Keywords (and commentary) must be separated by slashes (/)

- Keywords must contain at least one capital letter.

Follow these rules, and, not only will your `Command:` prompts look like AutoCAD `Command:` prompts, your shortcut menus will look like AutoCAD shortcut menus.

Inside Visual LISP, create (or load) the program shown in Listing 2–26.

Listing 2–26

```
;;; Filename: AP-02-26.lsp
(defun get-circ-type ()
  (initget "3P 2P Ttr")                      ; establish keywords
  (ap-getpoint                               ; get point or keyword
    nil
    (strcat "\nSpecify center point for circle"
            " or [3P/2P/Ttr (tan tan radius)]: "
    )
  )
)
```

Here we have a code fragment that emulates the first prompt from the CIRCLE command. The **initget** function specifies the keywords and abbreviations we use, and the prompt passed to the **ap-getpoint** function is the standard prompt for the CIRCLE command.

Again, we've used the **ap-getpoint** function because it updates the LASTPOINT system variable.

Select Tools|Load Text in Editor.

Enter the following at the AutoCAD Command: prompt:

```
Command: (get-circ-type)
Specify center point for circle or [3P/2P/Ttr (tan tan radius)]:
```
(Right-click on graphics area. Displays same shortcut menu as CIRCLE *command, as shown in Figure 2–2.)* .

```
Specify center point for circle or [3P/2P/Ttr (tan tan radius)]:
    2P (Pick the 2P option, and the menu types it for you.)
    "2P"
```

That was easy.

SUMMARY

In this chapter, we've discovered how easy but nonintuitive it is to create user interfaces that emulate AutoCAD's intrinsic commands. To this end, we learned how to use the Artful Programming API functions to

- Eliminate extraneous command line echoing.

- Restore AutoCAD system variables when our commands exit.

- Honor the U and UNDO commands.

- Restore AutoCAD system variables when the user cancels the command.

- Prevent transparent evocation of nontransparent routines

- Create a program skeleton with which to develop new programs.

- Protect the **command** and **vl-cmdf** functions from system variable settings.

- Respect the LASTPOINT system variable.

- Honor noun–verb object selection

- Honor the previous selection set.

- Immunize our program from UCS settings.

- Use selection set filters to restrict user inputs.

- Validate and restrict our user inputs.

- Provide default values for user prompts and save them with the drawing file.

- Honor context menus.

CHAPTER 3

Managing Dialog Boxes

The modal dialog box has lately become the preferred (and in some cases, only) paradigm for user interaction with AutoCAD; if you want the command line versions of the LAYER, LINETYPE, OSNAP, STYLE, or XREF commands, you must prefix the commands with a hyphen (-). Modal dialog boxes are distinguished from nonmodal dialog boxes insofar as they prevent further user interaction with a program until the dialog box is dismissed.

Functions for displaying modal dialog boxes and processing user input are provided in AutoLISP and Object ARX, while the dialog boxes themselves are defined in AutoCAD's Dialog Control Language (DCL). Together, they allow you to design and implement your own modal dialog boxes, similar to the ones used by native AutoCAD commands.

Despite Autodesk's attempt to have us call this capability PDB (Programmable Dialog Box)—and despite the fact that the Dialog Control Language is only *part* of PDB— we developers still refer to the design and implementation of dialog box–based commands as DCL programming.

In the discussion that follows, it is assumed that you have already been somewhat successful with DCL programming and that you want to learn more ways to work the widgets (as the building blocks of dialog boxes are sometimes called).

KEEPING PUT

Windows conveniently lets users move dialog boxes, but most of AutoCAD's commands insist on placing the dialog box squarely in the middle of our screens, ignoring our attempts to move (and keep) them out of the way. Here's how you can have your dialog boxes stay put.

Let's look at a minimal program to display a modal dialog box, as shown in Figure 3–1. The DCL code is shown in Listing 3–1. The AutoLISP code that drives it is shown in Listing 3–2.

Figure 3–1

Listing 3–1

```
// Filename: AP-03-01.dcl
hello: dialog {
  label = "Hello, World";
  : text { label = "Hello, World."; }
  ok_only;
}
```

Listing 3–2

```
;;; Filename: AP-03-02.lsp
(defun c:hello (/ dcl_id)
  (setq dcl_id (load_dialog "ap-03-01.dcl"))     ; Load the DCL file
  (if (not (new_dialog "hello" dcl_id))          ; Display the dialog
    (exit)
  )
  (start_dialog)                                 ; Wait until done
  (princ)                                        ; Exit quietly
)
```

Load and run your program, and you should find the dialog box centered on the screen, as shown in Figure 3–2.

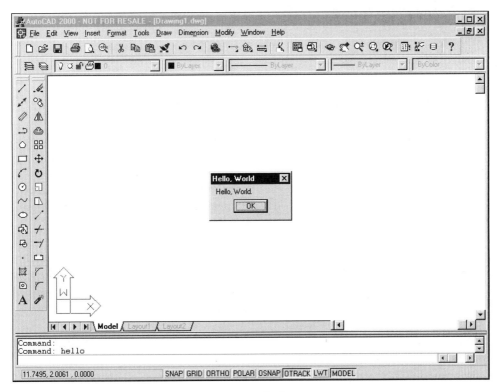

Figure 3–2

While the dialog box is still active, move it to some other position by dragging the title bar and click on the OK button.

Each time you evoke the dialog box (by typing **hello** at the Command: prompt) it starts right where it wants to—specifically, the center of the screen.

Listing 3–3 shows the few changes we have to make to our original program file to keep track of where we left things.

Listing 3–3

```
;;; Filename: AP-03-03.lsp
(defun c:hello (/ dcl_id)
  (setq dcl_id (ap-load-dialog "ap-03-01.dcl"))  ; Load the DCL file
  (if (not (ap-new-dialog "hello" dcl_id))        ; display the dialog
    (exit)
  )
```

```
(action_tile "accept" "(ap-done-dialog 1)")    ; define action
(start_dialog)                                  ; wait until done
(princ)                                         ; exit quietly
)
```

ap-load-dialog, from the Artful Programming API, replaces **load_dialog**; if a DCL file is loaded, we don't bother reloading it. This speeds up subsequent invocations. This also allows **ap-new-dialog** and **ap-done-dialog** to keep track of dialog boxes by their dcl_id's.

ap-new-dialog, also from the Artful Programming API, replaces **new_dialog**. This function displays the dialog box in its last known location. If no location is known, the dialog box is displayed at its default location.

ap-done-dialog, again from the Artful Programming API, replaces every implicit or explicit **done_dialog**, noting the location of the dialog box at the time it's dismissed. Load and run this version of the program, and you'll find it remembers the last location of the dialog box each time it's run.

(In case you're wondering why I substituted dashes for underscores in the function names, let's just say I'm lazy and don't care to press the SHIFT key when typing, plus I never remember which character to use, so I just made them all dashes.)

YOU SHOULD BE COMMITTED

Primates learn by experimentation. You may invoke a command (perhaps unintentionally, perhaps to see what it might do), which, in turn, invokes a dialog box. As long as the dialog box is equipped with both Accept and Cancel buttons, you should feel free to play with the dialog box with the expectation that if you click the Cancel button, everything will be undone. Only by pressing that Accept button do you commit to action.

From a programmer's point of view, the dialog box allows you to interactively manipulate some parameters, in preparation of performing some action based on these parameters, should you commit to the action by clicking the Accept button. Should you click the Cancel button, the parameters are to be restored their prior states.

Thus, from both the users' and the programmer's points of view, clicking Cancel not only cancels the command but undoes anything that was done prior to clicking it. Here is a way to cleanly cancel by using the Artful Programming API.

Listing 3–4 shows the DCL code for the dialog box as shown in Figure 3–3. The AutoLISP code that drives it is shown in Listing 3–5.

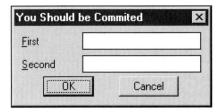

Figure 3–3

Listing 3–4

```
// Filename: AP-03-04.dcl
commit: dialog {
  label = "You Should be Committed";
  : edit_box {
     label      = "&First";
     key        = "first";
     edit_width = 20;
  }
  : edit_box {
     label      = "&Second";
     key        = "second";
     edit_width = 20;
  }
  ok_cancel;
}
```

Listing 3–5

```
;;; Filename: AP-03-05.lsp
(defun c:commit (/ dcl_id defaults params saved_params what_next)
  (setq defaults '((first . "") (second . "")))  ; parameters to control
  (setq params (mapcar 'car defaults))
  (setq dcl_id (ap-load-dialog "ap-03-04.dcl"))  ; load the dcl file
  (ap-init-vars defaults)                        ; initialize parameters
  (setq saved_params (ap-save-vars params))      ; save parameters
  (if (not (ap-new-dialog "commit" dcl_id))      ; display the dialog
    (exit)
  )
```

```
  (ap-set-tiles params)                        ; initialize tiles
  (ap-action-tiles params "callback_")         ; set actions
  (action_tile "accept" "(ap-done-dialog 1)")  ; set accept action
  (setq what_next (start_dialog))              ; wait until done
  (cond ((= 0 what_next)                        ; 'cancel' was pressed
         (ap-reset-vars saved_params)          ; reset the parameters
         (princ "Cancelled")                   ; print a message
         )
        ((= 1 what_next)                        ; 'ok' was pressed
         (princ "Parameters saved")            ; print a message
         )
     )
   (princ)                                      ; exit quietly
 )
(defun callback_first (key value) (set (read key) value))
(defun callback_second (key value) (set (read key) value))
```

The two minimalist callback functions, **callback_first**, and
callback_second, set the variables `first` and `second` to the contents of the
similarly named edit boxes whenever the specified edit box loses focus. Thus, the
variables automatically track the same-named tiles in the dialog box.

The `params` variable is a list of the parameters (AutoLISP variables) we want to be
controlled by the dialog box.

The **ap-init-vars** function, from the Artful Programming API, ensures every
listed parameter, if null, is set to the desired default value.

The **ap-save-vars** function, from the Artful Programming API, returns a dotted-
pair list of parameters and values. This will be used to restore values in the event the
dialog box is cancelled.

The **ap-set-tiles** function, from the Artful Programming API, sets each
tile to the identically named parameter. In this case, we've linked the AutoLISP
variables `first` and `second` to the tiles whose keys are "`first`" and
"`second`" respectively.

Tip: The Artful Programming API requires that the key value for each tile be the *lowercase* name of the AutoLISP variable to be controlled.

We use the **ap-action-tiles** function, from the Artful Programming API, to connect each tile to its respective callback function. Thus the "first" tile is connected to the **callback_first** function, and the "second" tile is connected to the **callback_second** function.

The **ap-reset-vars** function, from the Artful Programming API, accepts the list of dotted pairs created by **ap-save-params** and resets each parameter to its saved value.

 Tip: If you have multiple dialog boxes that reference the same variable name but with different callback functions, be sure to specify a unique prefix for the ap-action-tiles function.

NESTING IS NOT JUST FOR BIRDS

Dialog boxes are like rooms. Moving a user from one dialog box to another just to answer some question is like moving you from the dining room to the kitchen just to ask if you want a drink of water.

Nested dialog boxes should be avoided whenever possible. Nevertheless, the time will come when you must launch another dialog box, get some inputs, and return to the first one. Here's how it's done.

Listing 3–6 shows the DCL code for the nested dialog boxes shown in Figure 3–4. The AutoLISP code that drives it is shown in Listing 3–7.

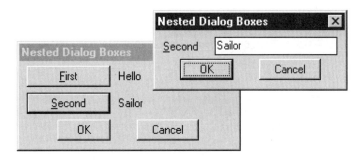

Figure 3–4

Listing 3–6

```
// Filename: AP-03-06.dcl
nested : dialog {
  label = "Nested Dialog Boxes";
   : row {
     : button { label = "&First"; key = "nested_1"; width = 15;}
      : text   { label = "";       key = "first";    width = 20;}
   }
   : row {
     : button { label = "&Second"; key = "nested_2"; width = 15;}
      : text   { label = "";         key = "second";   width = 20;}
   }
  ok_cancel;
}
nested_1 : dialog {
  label = "Nested Dialog Boxes";
  : edit_box { label = "&First"; key = "first"; edit_width = 20; }
  ok_cancel;
}
nested_2 : dialog {
  label = "Nested Dialog Boxes";
  : edit_box { label = "&Second"; key = "second"; edit_width = 20; }
  ok_cancel;
}
```

Listing 3–7

```
;;; Filename: AP-03-07.lsp
(defun c:nested (/ dcl_id what_next nest_buttons defaults params)
  (setq dcl_id (ap-load-dialog "ap-03-06.dcl"))
  (setq defaults '((first . "") (second . "")))
  (setq params (mapcar 'car defaults))
  (setq nest_buttons '("nested_1" "nested_2"))
  (ap-init-vars defaults)
  (setq what_next (nest "nested" dcl_id params nest_buttons))
  (if (= what_next 1)
    (princ "\nBingo!")
  )
  (princ)
)
```

```
(defun callback_first (key value) (set (read key) value))
(defun callback_second (key value) (set (read key) value))
(defun nest (dcl_name dcl_id params nest_buttons / what_next
            saved_params)
  (setq saved_params (ap-save-vars params))      ; save parameters
  (if (not (ap-new-dialog dcl_name dcl_id))      ; display the dialog
    (exit)
  )
  (ap-set-tiles params)                          ; initialize tiles
  (ap-action-tiles params "callback_")           ; set actions
  (nesting-buttons nest_buttons)                 ; set buttons
  (action_tile "accept" "(ap-done-dialog 1)")    ; set ok button
  (setq what_next (start_dialog))                ; wait until done
  (if (= 0 what_next)                            ; 'cancel' was pressed
    (ap-reset-vars saved_params)                 ; reset the parameters
  )
 what_next
)
(defun nesting-buttons (buttons)
  (foreach button buttons
    (action_tile
      button
      (strcat "(nest $key dcl_id params nest_buttons)"
                                                 ; pop up the dialog
          "(ap-set-tiles params)"                ; repaint the tiles
      )
    )
  )
)
```

Functionally, the code is very similar to that described in the previous section. Of special interest here is the action assigned to each of the nest buttons. Specifically, each nest button nests the dialog box with the same key as the nest button. When the nested dialog box is dismissed, the tiles in the previously topmost dialog box are repainted, just in case the nested box changed any parameters.

Especially noteworthy should be the code's utter lack of concern as to where a particular nest button or edit box is located. This allows you to redefine the content of nested dialog boxes by simply rearranging the DCL code; the AutoLISP code remains unchanged.

NOWHERE TO HIDE

Like it or not, when a dialog box is displayed, AutoCAD will not let you pick, point, or select objects. In fact, it won't let you perform any command or function that (even potentially) modifies the (potentially underlying) graphics screen. If you want to do so, you must temporarily hide any active dialog boxes. AutoCAD won't do this for you, but the Artful Programming API makes it simple.

Listing 3–8 shows the DCL code for the dialog box as shown in Figure 3–5. The AutoLISP code that drives it is shown in Listing 3–9.

Figure 3–5

Listing 3–8

```
// Filename: AP-03-08.dcl
hidebox: dialog {
  label = "Nowhere to Hide";
  : row {
    : button {key = "select"; label = "< &Select";}
    : text { key = "selected"; width = 10; }
  }
  ok_only;
}
```

Listing 3–9

```
;;; Filename: AP-03-09.lsp
(defun c:hidebox (/ dcl_id what_next ss1)
  (setq dcl_id (ap-load-dialog "ap-03-08.dcl"))  ; load the dcl file
  (setq what_next 999)
  (while (> what_next 1)                          ; wait until done
    (if (not (ap-new-dialog "hidebox" dcl_id))    ; display the dialog
      (exit)
    )
```

```
(set_tile "selected"
       (strcat (itoa (ap-sslength ss1)) " selected.")
 )
(action_tile "accept" "(ap-done-dialog 1)")
(action_tile "select" "(ap-done-dialog 2)")

(setq what_next (start_dialog))

(cond ((= 2 what_next)                        ; 'select' was pressed
       (setq ss1 nil)
       (setq ss1 (ssget))
       )
      ((= 1 what_next)                        ; 'ok' was pressed
       (princ "\nBingo!")
       )
   )
 )
(princ)                                       ; exit quietly
 )
```

First let's consider the code within the **while** loop. Each time the loop is executed:

- The dialog box is displayed.
- A callback function is defined for the Select button such that the subsequent **start_dialog** function call returns a value of 2.
- The **cond** function determines the action to be taken based on the value returned from the **start_dialog** function call.

The **while** loop repeats until the user responds with OK or Cancel.

Keep in mind that this example is incomplete. Having selected some items, you probably want to do *something* with the selection set other than printing "Bingo!"

A user-defined ***error*** function should be employed to avoid an

```
; error: Function cancelled
```

message should the user press ESC while selecting objects.

A ROOM WITH A VIEW

The Artful Programming Utilities make it simple to display AutoCAD slides in your dialog boxes.

Listing 3–10 shows the DCL code for the dialog box shown in Figure 3–6. The AutoLISP code that drives it is shown in Listing 3–11.

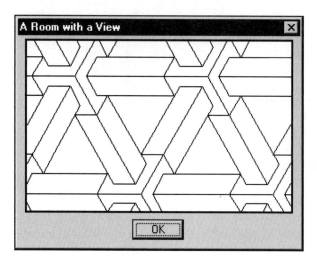

Figure 3–6

Listing 3–10

```
// Filename: AP-03-10.dcl
vslide: dialog {
  label = "A Room with a View";
  : image {
     key = "room";
     aspect_ratio = 0.63;
     width =        50;
     border =        2;
     border_color = -18;
     background =    0;
  }
  ok_only;
}
```

Listing 3–11

```
;;; Filename: AP-03-11.lsp
(defun c:viewslide (/ dcl_id)
  (setq dcl_id (ap-load-dialog "ap-03-10.dcl"))  ; load the dcl file
  (if (not (ap-new-dialog "vslide" dcl_id))      ; display the dialog
    (exit)
  )
  (action_tile "accept" "(ap-done-dialog 1)")    ; define action
  (ap-vslide "room" "acad(escher)")              ; view the slide
  (start_dialog)                                 ; wait until done
  (princ)                                        ; exit quietly
)
```

This simple routine displays a slide in the dialog box. The **ap-vslide** function from the Artful Programming API paints the specified image tile or image button with the specified slide; in this case room and acad(escher), respectively.

Of special interest to the artful programmer is the use of three user-defined tile attributes in the DCL code, shown in Figure 3–6: border, border_color, and background. The appearance of a particular slide depends on these three attributes.

The *border* attribute specifies the width of a frame (in pixels) to be drawn about the slide. The default is to display no border (*border* = *0*).

The *border_color* and *background* attributes specify the colors for the border and the slide background, respectively. They default to the current background of the AutoCAD graphics screen.

Colors may be specified with AutoCAD color numbers (ACI) or by one of the logical color numbers shown in Table 3–1.

Table 3–1: Image Tile Colors

Color number	Meaning
0	Current background of the AutoCAD graphics screen. Slides will appear in the dialog box just as they do with the VSLIDE command.
−15	Current dialog box background color.
−16	Current dialog box foreground color (for text).
−18	Current dialog box line color.

Before displaying the slide, the specified tile is filled with the background color specified in the DCL file, and a border of the width specified in the DCL file is drawn around the slide area.

Tips: Although you may specify an independent slide file, for best performance, you should specify a slide in a slide library.

For one way to create a slide library, access AutoCAD 2000 Help, select the Index tab, and enter **slidelib** as the word for which you are looking.

STRING ME ALONG

Although conceptually simple from the user's point of view, implementing edit boxes can be challenging for the artful programmer.

If we are to follow the artful prime directive (no AutoCAD program should ever do anything an AutoCAD user does not reasonably expect it to do), the artful programmer is expected to:

- Remove leading and trailing spaces from user inputs.
- Allow/disallow null (empty) values.
- Allow/disallow spaces.
- Provide meaningful error messages as required.

Fortunately, the artful programmer has access to the Artful Programming API, which, as you would hope, makes short shrift of the task by employing custom tile attributes.

Listing 3–12 shows the DCL code for the dialog box shown in Figure 3–7. The AutoLISP code that drives it is shown in Listing 3–13.

Figure 3–7

Listing 3–12

```
// Filename: AP-03-12.dcl
getstrings : dialog {
  label = "What's Your Name?";
  : spacer { height = 0.5; }
  : edit_box {
    label      = "&First";
    key        = "first_name";
    edit_width = 20;
    tag        = "First Name";
    trim       = true;
  }
  : edit_box {
    label      = "&Middle";
    key        = "middle_name";
    edit_width = 20;
    tag        = "Middle Name";
    trim       = true;
    null_ok    = true;
  }
  : edit_box {
    label      = "&Last";
    key        = "last_name";
    edit_width = 20;
    tag        = "Last Name";
    trim       = true;
    spaces_ok  = true;
  }
  : spacer { height = 0.5; }
  ok_cancel_err;
}
```

Listing 3–13

```
;;; Filename: AP-03-13.lsp
(defun c:getstrings (/ dcl_id defaults params saved_params what_next)
  (setq
    defaults '((first_name . "") (middle_name . "") (last_name . ""))
  )
  (setq params (mapcar 'car defaults))
```

```
(setq dcl_id (ap-load-dialog "ap-03-12.dcl"))      ; load the dcl file
(ap-init-vars defaults)                            ; initialize parameters
(setq saved_params (ap-save-vars params))          ; save parameters
(if (not (ap-new-dialog "getstrings" dcl_id))      ; display the dialog
  (exit)
)
(ap-set-tiles params)                              ; initialize tiles
(ap-action-tiles params "callback_")               ; set actions
(action_tile
  "accept"
  "(ap-done-dialog-conditional \"callback_\" params 1)"
  )                                                ; set accept action
(setq what_next (start_dialog))                    ; wait until done
(cond ((= 0 what_next)                             ; 'cancel' was pressed
       (ap-reset-vars saved_params)                ; reset the parameters
       )
      ((= 1 what_next)                             ; 'ok' was pressed
       (princ "\nBingo!")
       )
 )
 (princ)                                           ; exit quietly
)
(defun callback_first_name (key value)
  (ap-callback-string key value)
)
(defun callback_middle_name (key value)
  (ap-callback-string key value)
)
(defun callback_last_name (key value)
  (ap-callback-string key value)
)
```

Our goal here is simple. The dialog box shown in Figure 3–7 controls three parameters, first_name, middle_name, and last_name, via the three corresponding edit boxes.

If the user clicks the OK button, the program stores the values entered into the dialog box. If the user cancels, no action is taken. What's new here?

The **ap-done-dialog-conditional** function, from the Artful Programming API, performs in turn each of the callback functions defined with **ap-action-tiles**. If any callback fails (returns nil for whatever reason), **ap-done-dialog-conditional** sets the focus to the tested tile. If no callback fails, **ap-done-dialog-conditional** performs an **ap-done-dialog**, closing the dialog box.

By using **ap-done-dialog-conditional** to close our dialog box, we are thus assured we won't get an OK if the user has made an invalid entry.

How do we know if the user has made an invalid entry? It's up to each and every callback function to return a non-nil value if the user input is valid, and a nil value if it's invalid.

Tip: Every callback function must return a non-nil value if the user input is valid, and a nil value if it is not.

What else is new here?

The **ap-callback-string** function from the Artful Programming API handles trimming and error checking specific to the edit box that invoked it. The **ap-callback-string** function uses the custom tile attributes shown in Table 3–2 to determine the trimming and error checks to be performed.

Table 3–2: Custom Tile Attributes—ap-callback-string

Attribute	Usage
tag	The description of the tile to appear in error messages.
trim	Set to true to trim leading and trailing spaces.
null_ok	Set to true to accept blank entries.
spaces_ok	Set to true to accept internal spaces.

The DCL code shown in Listing 3–12 instructs **ap-callback-string** as shown in Table 3–3.

Table 3–3: What's Your Name? Input Validation

Key	ap-callback-string Actions
first_name	Trim the string Disallow an empty string Disallow spaces in the string
middle_name	Trim the string Allow an empty string Disallow spaces in the string
last_name	Trim the string Disallow an empty string Allow spaces in the string

If any of these rules is not met, the **ap-callback-string** function displays an error message and returns nil. If all of the rules are met, **ap-callback-string** sets the variable whose name is the key of the tile.

To add another string edit box to this application,

- Add the variable to be controlled, with an optional default value, to the variable defaults.

- Add a callback function for this variable.

- Add the edit box, with appropriate attributes, to the DCL code.

That's all you have to do. Pretty cool, huh?

GETTING INTEGERS

Accepting integers from an edit box comes with its own requirements. Again, edit boxes allow the user to type any darned thing they please.

If we are to follow the artful prime directive again, the artful programmer is expected to:

- Allow/disallow null (empty) values.

- Allow/disallow negative values.

- Allow/disallow zero values.

- Provide meaningful error messages as required.

The Artful Programming API makes it easy by employing custom tile attributes.

Listing 3–14 shows the DCL code for the dialog box shown in Figure 3–8. The AutoLISP code that drives it is shown in Listing 3–15.

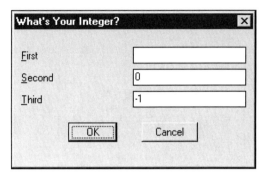

Figure 3–8

Listing 3–14

```
// Filename: AP-03-14.dcl
getints : dialog {
  width = 45;
  label = "What's Your Integer?";
  : spacer { height = 0.5; }
  : edit_box {
   label      = "&First";
   key        = "int1";
   edit_width = 20;
   tag        = "First Number";
  }
  : edit_box {
   label      = "&Second";
   key        = "int2";
   edit_width = 20;
   tag        = "Second Number";
   null_ok    = true;
   minus_ok   = true;
  }
  : edit_box {
   label      = "&Third";
```

```
  key        = "int3";
  edit_width = 20;
  tag        = "Third Number";
  zero_ok    = true;
  }
  : spacer { height = 0.5; }
  ok_cancel_err;
}
```

Listing 3–15

```lisp
;;; Filename: AP-03-15.lsp
(defun c:getints (/ dcl_id defaults params saved_params what_next)
  (setq defaults '((int1) (int2 . 0) (int3 . -1)))
  (setq params (mapcar 'car defaults))
  (setq dcl_id (ap-load-dialog "ap-03-14.dcl"))   ; load the dcl file
  (ap-init-vars defaults)                          ; initialize parameters
  (setq saved_params (ap-save-vars params))        ; save parameters
  (if (not (ap-new-dialog "getints" dcl_id))       ; display the dialog
    (exit)
  )
  (ap-set-tiles params)                            ; initialize tiles
  (ap-action-tiles params "callback_")             ; set actions
  (action_tile                                     ; set accept action
    "accept"
    "(ap-done-dialog-conditional \"callback_\" params 1)"
  )
  (setq what_next (start_dialog))                  ; wait until done
  (cond ((= 0 what_next)                           ; 'cancel' was pressed
         (ap-reset-vars saved_params)              ; reset the parameters
         )
        ((= 1 what_next)                           ; 'ok' was pressed
         (princ "\nBingo!")
         )
  )
  (princ)                                          ; exit quietly
)
(defun callback_int1 (key value)
  (ap-callback-int key value)
)
```

```
(defun callback_int2 (key value)
  (ap-callback-int key value)
)
(defun callback_int3 (key value)
  (ap-callback-int key value)
)
```

Our goal here is simple. The dialog box shown in Figure 3–8 controls three parameters, int1, int2, and int3, via the three corresponding edit boxes. If the user clicks on OK, the program stores the values entered into the dialog box. If the user cancels, no action is taken.

You will notice that little has changed from our previous example. All we have are some different variable names and some different callback functions.

The **ap-set-tiles** function, from the Artful Programming API, automatically displays integer values as strings.

The **ap-callback-int** function, from the Artful Programming API, handles error checking apropos of entering an integer. The custom tile attributes (shown in Table 3–4) determine the error checks to be performed.

Table 3–4: Custom Tile Attributes—ap-callback-int

Attribute	Usage
tag	The description of the tile to appear in error messages.
null_ok	Set to true to allow null values.
minus_ok	Set to true to accept negative values.
zero_ok	Set to true to accept zero values.

The DCL code shown in Listing 3–14 instructs **ap-callback-int**, as shown in Table 3–5.

Table 3–5: What's Your Integer? Input Validation

Key	ap-callback-int Actions
int1	Disallow null Disallow minus Disallow zero
int2	Allow null Allow minus Disallow zero
int3	Disallow null Disallow minus Allow zero

If any of these rules is not met, the **ap-callback-int** function displays an error message and returns nil. If all of the rules are met, **ap-callback-int** sets the variable whose name is the key of the tile to the integer value of the tile and returns the integer value of the variable, or "" if it's nil.

To add another integer edit box to this application:

- Add the variable to be controlled, with an optional default value, to the variable defaults.

- Add the callback function for this variable.

- Add the edit box with appropriate attributes to the DCL code.

What's the difference in coding a string edit box and an integer edit box? Just the callback function. That's all you have to do.

GETTING REALS

Accepting real numbers from an edit box comes with its own requirements. As with integers, the artful programmer must

- Allow/disallow null (empty) values.

- Allow/disallow negative values.

- Allow/disallow zero values.

- Provide meaningful error messages as required.

Just to complicate matters, real numbers in AutoLISP are used to represent angles (expressed in radians), distances, and, well, real numbers. Now you must deal with both input and output conversions (between the real number in the variable and the string value in the tile). The Artful Programming API makes it easy.

Listing 3–16 shows the DCL code for the dialog box shown in Figure 3–9. The AutoLISP code that drives it is shown in Listing 3–17.

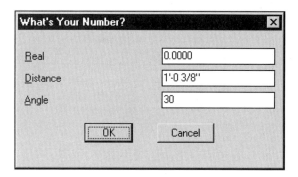

Figure 3–9

Listing 3–16

```
// Filename: AP-03-16.dcl
getreals : dialog {
  width = 50;
  label = "What's Your Number?";
  : spacer { height = 0.5; }
  : edit_box {
   label       = "&Real";
   key         = "real1";
   edit_width = 20;
   tag         = "Real";
   units       = 2;
   prec        = 4;
  }
  : edit_box {
   label       = "&Distance";
   key         = "real2";
   edit_width = 20;
   tag         = "Distance";
  }
  : edit_box {
```

```
 label       = "&Angle";
 key         = "real3";
 edit_width = 20;
 tag         = "Angle";
 is_angle    = true;
 minus_ok    = true;
 zero_ok     = true;
 }
 : spacer { height = 0.5; }
 ok_cancel_err;
}
```

Listing 3–17

```
;;; Filename: AP-03-17.lsp
(defun c:getreals (/ dcl_id defaults params saved_params what_next)
  (setq defaults '((real1 . 0.0) (real2 . 12.375) (real3 . 0.523599)))
  (setq params (mapcar 'car defaults))
  (setq dcl_id (ap-load-dialog "ap-03-16.dcl"))   ; load the dcl file
  (ap-init-vars defaults)                          ; initialize parameters
  (setq saved_params (ap-save-vars params))        ; save parameters
  (if (not (ap-new-dialog "getreals" dcl_id))      ; display the dialog
    (exit)
  )
  (ap-set-tiles params)                            ; initialize tiles
  (ap-action-tiles params "callback_")             ; set actions
  (action_tile                                     ; set accept action
    "accept"
    "(ap-done-dialog-conditional \"callback_\" params 1)"
  )
  (setq what_next (start_dialog))                  ; wait until done
  (cond ((= 0 what_next)                           ; 'cancel' was pressed
         (ap-reset-vars saved_params)              ; reset the parameters
         )
        ((= 1 what_next)                           ; 'ok' was pressed
         (princ "\nBingo!")
         )
  )
  (princ)                                          ; exit quietly
)
```

```
(defun callback_real1 (key value)
  (ap-callback-real key value)
)
(defun callback_real2 (key value)
  (ap-callback-real key value)
)
(defun callback_real3 (key value)
  (ap-callback-real key value)
)
```

Our goal, again, is simple. The dialog box shown in Figure 3–9 controls three parameters, real1, real2, and real3, via the three corresponding edit boxes. If the user clicks on OK, the program stores the values entered via the dialog box. If the user cancels, no action is taken.

Again, little has changed from our previous example. We have some different variable names and some different callback functions, though.

The **ap-callback-real** function from the Artful Programming API handles error checking specific to the edit box that invoked it. The custom tile attributes shown in Table 3–6 determine the error checks and input conversion to be performed.

The **ap-set-tiles** function from the Artful Programming API formats numbers according to values of the custom tile attributes is_angle, units, and prec.

Table 3–6: Custom Tile Attributes—ap-callback-real

Attribute	Usage
tag	The description of the tile to appear in error messages.
null_ok	Set to true to allow null values.
minus_ok	Set to true to accept negative values.
zero_ok	Set to true to accept zero values.
is_angle	Set to true to control angles expressed in radians.
units	Controls units display of numbers. Defaults to LUNITS or AUNITS.
prec	Controls precision display of numbers. Defaults to LUPREC or AUPREC.

By using the custom tile attributes as shown in Listing 3–16, we've instructed **ap-callback-real** to:

- Display and accept `real1` as a positive, nonzero fixed-point number with four decimal places.

- Display and accept `real2` as a positive, nonzero distance using the current drawing units settings.

- Display and accept `real3` as an angle using the current drawing units settings.

- Display an error message whenever one of the specified tests fails.

If all of the rules are met, **ap-callback-real** sets the variable whose name is the key of the tile to the real value of the tile and returns the real value of the variable, or "" if it's `nil`.

To add another real edit box to this application:

- Add the variable to be controlled, with an optional default value, to the variable `defaults`.

- Add the callback function for this variable.

- Add the edit box, with appropriate attributes, to the DCL code.

Again, it's all in the callback functions.

TOGGLES AND RADIO BUTTONS

Toggle buttons and radio groups are easily handled by the Artful Programming API.

Listing 3–18 shows the DCL code for the dialog box shown in Figure 3–10. The AutoLISP code that drives it is shown in Listing 3–19.

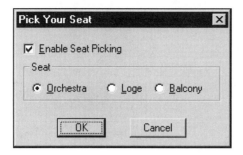

Figure 3–10

Listing 3–18

```
// Filename: AP-03-18.dcl
toggles : dialog {
 label = "Pick Your Seat";
 : spacer { height = 0.5; }
 : toggle {
  label     = "&Enable Seat Picking";
  key       = "toggle1";
 }
 : boxed_radio_row {
  label   = "Seat";
  key     = "radio1";
  buttons = "orchestra loge balcony";
   : radio_button {
      key   = "radio1_orchestra";
      label = "&Orchestra";
   }
   : radio_button {
      key   = "radio1_loge";
      label = "&Loge";
   }
   : radio_button {
      key   = "radio1_balcony";
      label = "&Balcony";
   }
 }
 : spacer { height = 0.5; }
 ok_cancel;
}
```

Listing 3–19

```
;;; Filename: AP-03-19.lsp
(defun c:toggles (/ dcl_id defaults params saved_params what_next)
  (setq defaults '((toggle1 . "1") (radio1)))
  (setq params (mapcar 'car defaults))
  (setq dcl_id (ap-load-dialog "ap-03-18.dcl"))  ; load the dcl file
  (ap-init-vars defaults)                         ; initialize parameters
  (setq saved_params (ap-save-vars params))       ; save parameters
  (if (not (ap-new-dialog "toggles" dcl_id))      ; display the dialog
    (exit)
  )
  (ap-set-tiles params)                           ; initialize tiles
  (enable_radio1)                                 ; enable/disable buttons
  (ap-action-tiles params "callback_")            ; set actions
  (action_tile                                    ; set accept action
    "accept"
    "(ap-done-dialog-conditional \"callback_\" params 1)"
  )
  (setq what_next (start_dialog))                 ; wait until done
  (cond ((= 0 what_next)                          ; 'cancel' was pressed
         (ap-reset-vars saved_params)             ; reset the parameters
         )
        ((= 1 what_next)                          ; 'ok' was pressed
         (princ "\nBingo!")
         )
  )
  (princ)                                         ; exit quietly
)
(defun callback_toggle1 (key value)
  (ap-callback-toggle key value)
  (enable_radio1)
)
(defun callback_radio1 (key value)
  (ap-callback-radio key value)
)
(defun enable_radio1 ()
  (mode_tile "radio1" (- 1 (atoi toggle1)))
  toggle1
)
```

The dialog box shown in Figure 3–10 controls two variables, `toggle1` and `radio1`.

The **ap-callback-radio** function from the Artful Programming API makes use of a custom tile attribute, along with a naming convention, to greatly simplify the programming of radio rows and columns. This is shown in Table 3–7.

Table 3–7: Custom Tile Attributes—ap-callback-radio

Attribute	Usage
buttons	A string specifying the values to be assigned by the radio buttons in the radio row or column, as in `"value1 value2 value3"`
key	Each of the buttons in a radio row or column must have a key of `varname_value`, where • `varname` is both the name of the radio row or column, *and* the name of the variable to be controlled • `value` is the value to be assigned to *varname* whenever said radio button is pushed.

The **ap-callback-toggle** function, from the Artful Programming API, sets the variable whose name is the key of the tile to a string value of "0" or "1," depending on the state of the toggle, and returns this value.

The **ap-set-tiles** function, from the Artful Programming API, sets the toggle corresponding to the current value of the string variable being controlled (in this case, `toggle1`). A value of "1" indicates the toggle is set.

The **ap-callback-radio** function, also from the Artful Programming API, sets the variable whose name is the key of the radio group to a string value corresponding to the key of the button pressed.

The **ap-set-tiles** function, yet again from the Artful Programming API, sets the radio button corresponding to the current value of the string variable being controlled (in this case, `radio1`). If there is no corresponding button or if the variable being controlled is `nil`, **ap-set-tiles** sets the variable to the key of the first radio button.

Referring to the DCL code in Listing 3–18, in this example we've instructed **ap-callback-radio** that pressing:

- `radio1_orchestra` sets `radio1` **to** "orchestra".
- `radio1_loge` sets `radio1` **to** "loge".
- `radio1_balcony` sets `radio1` **to** "balcony".

Note that we've added some extra code to the **callback_toggle1** function such that the toggle enables/disables the radio1 row. Remember that for **ap-done-dialog-conditional** to accept the dialog box, all callback functions must return a non-nil value if their corresponding values are properly set.

To add another toggle or radio group to this application,

- Add the variable to be controlled, with an optional default value, to the variable defaults.

- Add the callback function for this variable.

- Add the toggle or radio group, with appropriate attributes, to the DCL code. Do not forget the buttons attribute.

Easy, isn't it? It's all in the callback functions.

LIST BOXES AND POP-UP LISTS

As you might well assume, the Artful Programming API also makes it easy to program list boxes and pop-up lists.

Listing 3–20 shows the DCL code for the dialog box shown in Figure 3–11. The AutoLISP code that drives it is shown in Listing 3–21.

Figure 3–11

Listing 3–20

```
// Filename: AP-03-20.dcl
lists : dialog {
  label = "Pick Your List";
  : spacer { height = 0.5; }
  : popup_list {
      key   = "list1";
      label = "&Length Units";
      is_list = true;
  }
  : row {
    width = 35;
      : list_box {
        key   = "list2";
        label = "L&anguage";
        height = 6;
        is_list = true;
        sort    = true;
      }
      : list_box {
        key   = "list3";
        label = "&Shopping";
        height = 6;
        is_list = true;
        multiple_select = true;
      }
  }
  : spacer { height = 0.5; }
  ok_cancel;
}
```

Listing 3–21

```
;;; Filename: AP-03-21.lsp
(defun c:lists (/ dcl_id defaults params saved_params what_next)
  (setq defaults '((list1) (list2) (list3)))
  (setq params (mapcar 'car defaults))
  (setq list1_values
        '((1 . "Scientific")
```

```
                (2 . "Decimal")
                (3 . "Engineering")
                (4 . "Architectural")
                (5 . "Fractional")
                )
    )
    (setq list2_values '("APL" "FORTRAN" "BASIC" "AutoLISP" "C++"))
    (setq list3_values
            '("Fruits"        "Vegetables"    "Cereals"
            "Breads"        "Cookies"       "Potato Chips"
            )
    )
    (setq dcl_id (ap-load-dialog "ap-03-20.dcl"))   ; load the dcl file
    (ap-init-vars defaults)                         ; initialize parameters
    (setq list1 (list (getvar "lunits")))           ; initialize list1
    (setq saved_params (ap-save-vars params))       ; save parameters
    (if (not (ap-new-dialog "lists" dcl_id))        ; display the dialog
      (exit)
    )
    (ap-set-tiles params)                           ; initialize tiles
    (ap-action-tiles params "callback_")            ; set actions
    (action_tile                                    ; set accept action
      "accept"
      "(ap-done-dialog-conditional \"callback_\" params 1)"
    )
    (setq what_next (start_dialog))                 ; wait until done
    (cond ((= 0 what_next)                          ; 'cancel' was pressed
          (ap-reset-vars saved_params)             ; reset the parameters
          )
          ((= 1 what_next)                          ; 'ok' was pressed
          (princ "\nBingo!")
          (setvar "lunits" (car list1))            ; set the lunits variable
          )
    )
    (princ)                                         ; exit quietly
  )
(defun callback_list1 (key value)
  (ap-callback-list key value)
)
```

```
(defun callback_list2 (key value)
  (ap-callback-list key value)
)
(defun callback_list3 (key value)
  (ap-callback-list key value)
)
```

The dialog box shown in Figure 3–11 controls three variables, `list1`, `list2`, and `list3`.

Each variable to be controlled by a list box or pop-up list must be provided with a corresponding variable containing a list of values; this list specifies those values that may assigned to the controlled variable.

If the variable is to be set to a string value identical to the displayed value, a list of strings will suffice. If, on the other hand, you wish to assign a value other than the displayed value, a dotted-pair list is used. The **car** of each item in the list is the value to be assigned, whereas the **cdr** of each item is the display value.

In this example, `list1_values` contains the values that may be assigned to `list1`, `list2_values` contains the values that may be assigned to `list2`, and `list3_values` to `list3`.

The variable `list1` will be set to the integer values 1 to 5, corresponding to the **car** of each of the `list1_values`.

The expression

```
(setq list1 (list (getvar "lunits")))
```

sets `list1` to the current value of the LUNITS system variable.

 Note: Variables controlled by list boxes and pop-up lists must be lists, even if only one value is acceptable for the variable.

The **ap-set-tile** function initializes each list box or pop-up list to the values in its corresponding `_list` variable. It uses the tile attributes shown in Table 3–8.

Table 3–8: Custom Tile Attributes—ap-callback-list

Attribute	Usage
`tag`	The description of the tile to appear in error messages.
`is_list`	Set to `true` to identify a widget as a list box or popup list.
`sort`	Set to `true` to sort alphabetically the list values.

In addition, if the standard `multiple_select` attribute is *not* set to `true`, and the controlled variable is not one of the allowed values, it is set to the first value in the (possibly sorted) list.

Referring to the DCL code in Listing 3–20, in this example we've instructed **ap-set-tile** to sort only `list2`.

The **ap-callback-list** function sets the variable whose name is the key of the list box or pop-up list to a list of values corresponding to the selected row(s) of the list box or pop-up list.

The expression

```
(setvar "lunits" (car list1))
```

sets the LUNITS system variable to the current value of `list1`.

To add another list box or pop-up list to this application,

- Add the variable to be controlled with an optional default value to the variable `defaults`.
- Add the callback function for this variable.
- Add the list box or pop-up list, with appropriate attributes, to the DCL code. Do not forget the `is_list` attribute.

SUMMARY

In this chapter, we've learned how the use the Artful Programming API to control dialog boxes and to link AutoLISP variables to widgets in a dialog box. We've learned how to:

- Keep dialog boxes where the user positions them.

- Discard all variable changes if the dialog box is cancelled.

- Nest dialog boxes.

- Hide dialog boxes.

- Display slide images.

- Link string, integer, real, distance, and angle variables to edit boxes while validating and restricting user inputs.

- Link toggle variables to toggles.

- Link string variables to radio rows and columns.

- Link list variables to pop-up lists and list boxes.

- Validate all inputs before OKing a dialog box.

CHAPTER 4

Manipulating AutoCAD Drawings with ActiveX Automation

WHEN THE ONLY TOOL YOU HAVE IS A HAMMER, EVERYTHING LOOKS LIKE A NAIL

As an AutoCAD user, you've (presumably) created, queried, modified, and erased (that is, manipulated) graphical objects (*entities*) using at least some of AutoCAD's commands. You may, for example, have used the POLYGON command to draw a polyline, the ROTATE command to rotate it, the FILLET command to add some fillets, and the CHPROP command to modify its color.

You've also presumably used AutoCAD's commands to modify nongraphical objects as well. You may have used the LAYER command to manipulate layers, the STYLE command to manipulate text styles, and the DDIM command to manipulate dimension styles.

It's only natural, then, for the programs we write to use AutoCAD's commands as well. There's nothing wrong with this, and in many cases, this is the easiest (and certainly the most familiar) way for us to do so. This isn't, however, the only way to do so.

In this chapter, we'll review the AutoCAD DXF Model, introduce you to the AutoCAD Object Model, and show you how to create, query, modify, and delete AutoCAD graphical objects with ActiveX Automation, a technology developed by Microsoft and based on the Component Object Model (COM) architecture.

THE AUTOCAD DXF MODEL

The AutoCAD DXF Model enables the programmer to create, query, modify, and erase both graphical and non-graphical drawing data.

The AutoCAD DXF model is a representation of almost all the information contained in an AutoCAD drawing file. Five sections of the model are accessible via AutoLISP: header, tables, blocks, graphical objects (entities), and nongraphical objects. I've chosen this nomenclature to avoid (future) conflicts with the AutoCAD Object Model.

HEADER SECTION

This is where all the AutoCAD system variables that are stored with the drawing reside.

TABLES SECTION

The tables shown in Table 4–1 are somewhat accessible from AutoLISP.

Table 4–1: AutoCAD DXF Model Tables

AutoCAD Table	Function
Appid	The Application ID table contains the names and status of all registered applications.
Block	The Block table contains the names and status of the user blocks, anonymous blocks, and xrefs in the drawing, as well as pointers to their definitions in the blocks section.
Dimstyle	The Dimension Style table contains the dimension styles.
Layer	The Layer table contains the layer names and status.
LType	The LType table does it for line type definitions.
Style	The Style table contains the text style definitions.
UCS	The UCS table contains the UCS definitions.
View	The View table contains the named views in the drawing.
VPort	The Viewport table contains the viewport definitions.

BLOCKS SECTION

This section contains the graphical entities defining all the user blocks, anonymous blocks, and xrefs in the drawing.

GRAPHICAL ENTITIES SECTION

This section contains all the graphical entities in paper space and model space.

NONGRAPHICAL OBJECTS SECTION

Nongraphical objects are similar to graphical entities, except that they have no intrinsic graphical or geometric meaning—they are anything that didn't fit anywhere else the drawing. To quote the AutoCAD DXF Reference, "This section represents a homogeneous heap of objects with topological ordering of objects by ownership, such that the owners always appear before the objects they own."

AUTOLISP <-> AUTOCAD DATA FORMAT

To the AutoLISP programmer, every entry in every one of these tables is an association list similar to that returned by the **entget** function and accepted by the **entmake** and **entmod** functions. Let's give them a try:

Enter the following at the Visual LISP console:

```
_$ (vl-cmdf "circle" '(1 2) 3) (Draw a circle.)
T
_$ (setq Circle (entget (entlast)))
```

Here's what it looks like in the DXF model:

```
((-1 . <Entity name: 1b52d60>)
  (0 . "CIRCLE")
  (330 . <Entity name: 1b52cf8>)
  (5 . "2C")
  (100 . "AcDbEntity")
  (67 . 0)
  (410 . "Model")
  (8 . "0") (Believe it or not, after ten or so years of dealing with this, one learns that
        10 group is the center, the 40 group is the radius, and the 8 group is the layer.)
  (100 . "AcDbCircle") (I'd have to look up the other fields in the reference.)
  (10 1.0 2.0 0.0)
  (40 . 3.0)
  (210 0.0 0.0 1.0))
```

That wasn't all that bad. Now consider what you get when you examine a dimension style:

```
_$ (setq Dimstyle (entget (tblobjname "dimstyle" "standard")))
((-1 . <Entity name: 1b52d38>)
  (0 . "DIMSTYLE")
  (105 . "27")
  (330 . <Entity name: 1b52c50>)
  (100 . "AcDbSymbolTableRecord")
  (100 . "AcDbDimStyleTableRecord")
  (2 . "Standard")
  (70 . 0)
  (340 . <Entity name: 1b52c88>))
```

Now let's move on to the AutoCAD Object Model.

THE AUTOCAD OBJECT MODEL

As an AutoCAD user, you're already familiar with the terms *objects* and *properties* as they apply to graphical objects—such stuff as drawings are made of. In the world of ActiveX Automation, *everything is an object.*

The concept of objects is extended to both the graphical objects (lines, circles, arcs, etc.) and the nongraphical objects (layers, linetypes, dimstyles, etc.) that make up an AutoCAD drawing. AutoCAD drawings are considered objects, as are the AutoCAD menus, toolbars, and preferences, as well as the AutoCAD application itself.

Precisely *which* properties an object has depends on the type of object. A circle object has the properties of `Center`, `Radius`, `Length`, and `Area` (among others), whereas a line object has the properties of `StartPoint`, `EndPoint`, and `Length` (again, with others), but not `Radius` or `Center`.

A *method* is just the ActiveX Automation terminology for a function that may be applied to an object. Methods include editing commands, such as COPY, ROTATE3D, and ERASE.

In ActiveX Automation terminology, we utilize the methods and properties of an object to query and modify the object.

WHY USE ACTIVEX AUTOMATION?

Perhaps a better title for this section should be "Why Should I Learn Something New?" This is the very question I asked myself as I began to delve into the AutoCAD

Object Model. After all, I was writing good programs, within the limits of the AutoLISP environment, and my customers were happy.

Rather then spend the next thirty pages extolling the benefits of ActiveX Automation, let me just say this:

By using ActiveX Automation with AutoLISP, I've been able to write programs that

- Were faster than programs using the AutoCAD command throat.

- Were faster than programs using **entget** and **entmod**.

- Could directly access and control virtually every aspect of the AutoCAD applications.

- Could do much more than programs utilizing AutoLISP alone.

SEEK AND YE MAY FIND

You may not have realized it, but here's what AutoCAD goes through every time you start an AutoCAD command, be it via the keyboard, a script file, a menu, toolbar macro, or a Visual LISP or ObjectARX application.

- AutoCAD checks to see if the command name is an empty string. If it is, AutoCAD uses the name of the last command issued.

- AutoCAD searches for the command name in its list of intrinsic commands. If it's not there, the search continues.

   If the command name is in this list, but is not preceded by a period (.), AutoCAD searches for the command name in the list of commands that have been undefined via the UNDEFINE command. If the command has been undefined, the search continues.

   If the command hasn't been undefined, AutoCAD checks to see if there's any compelling reason the command can't be run; for example, you issue the MVIEW command while in model space.

   If there's no reason the command can't be run, AutoCAD runs the command.

- AutoCAD searches for the command name in the list of commands defined by device drivers. If it isn't in the list, the search continues.

- AutoCAD searches for the command name in the list of commands defined by the display driver. If it isn't in the list, the search continues.

- AutoCAD searches for the command name in the list of external commands defined in the *acad.pgp* file. If it isn't in the list, the search continues.

- AutoCAD searches for the command name in list of commands defined by Visual LISP and ObjectARX applications. If it isn't in the list, the search continues.

- AutoCAD searches for the command name in the list of system variables. If it isn't in the list, the search continues.

- AutoCAD searches for the command name in the list of command aliases defined in the *acad.pgp* file. If it's in this list, AutoCAD replaces the command name with the name to which it's aliased, and resumes the search back at the first step.

- If AutoCAD still hasn't found the command, the search ends with an "Unknown command" error message.

Though the time AutoCAD spends searching for a command name is small compared to the time it takes for you to type it at the command line, this time is not insignificant when your Visual LISP application issues many AutoCAD commands.

Imagine, if you will, a Visual LISP command that changes the scales of selected block insertions via AutoCAD's SCALE command. Change 20,000 block insertions, and AutoCAD searches 20,000 times for the SCALE command. From the point of view of a Visual LISP program, and from the point of view of your users, this is wasted time.

WHERE DO YOU DRAW THE LINE?

Let's start exploring the ActiveX Automation interface.

Enter the following at the Visual LISP _$ prompt:

```
_$ (apax-init) (Initializes Artful Programming ActiveX Variables.)
apAcadApp                  #<VLA-OBJECT IAcadApplication 00ac8928>
apActiveDimStyle           #<VLA-OBJECT IAcadDimStyle 01dab284>
apActiveDocument           #<VLA-OBJECT IAcadDocument 00b6cbe4>
apActiveLayer              #<VLA-OBJECT IAcadLayer 01dab1c4>
apActiveLayout             #<VLA-OBJECT IAcadLayout 01dab0c4>
apActiveLinetype           #<VLA-OBJECT IAcadLineType 01dab024>
apActiveSelectionSet       #<VLA-OBJECT IAcadSelectionSet
                              01daee64>
apActiveSpace              1
apActiveTextStyle          #<VLA-OBJECT IAcadTextStyle 01daec64>
```

```
apActiveUCS             #<VLA-OBJECT IAcadUCS 01db6954>
apActiveViewport        #<VLA-OBJECT IAcadViewport 01db6a54>
apApplPath              C:\\NT Program Files\\ACAD2000
apBlocks                #<VLA-OBJECT IAcadBlocks 01dae754>
apCaption               AutoCAD 2000 - NOT FOR RESALE
apDatabase              #<VLA-OBJECT IAcadDatabase 01dae814>
apDatabasePreferences   #<VLA-OBJECT IAcadDatabasePreferences
                           01dae7e4>
apDictionaries          #<VLA-OBJECT IAcadDictionaries
                           01dae644>
apDimStyles             #<VLA-OBJECT IAcadDimStyles 01dae5a4>
apDocuments             #<VLA-OBJECT IAcadDocuments 01dac8f0>
apElevationModelSpace   0.0
apElevationPaperSpace   0.0
apFullName              C:\Artful Programming\Artful
                           Programming.dwg
apFullName              C:\NT Program Files\ACAD2000\acad.exe
apGroups                #<VLA-OBJECT IAcadGroups 01dae404>
apHeight                776
apLayers                #<VLA-OBJECT IAcadLayers 01dae364>
apLayouts               #<VLA-OBJECT IAcadLayouts 01dae254>
apLinetypes             #<VLA-OBJECT IAcadLineTypes 01dae1b4>
apLocaleId              1033
apMenuBar               #<VLA-OBJECT IAcadMenuBar 01dacb24>
apMenuGroups            #<VLA-OBJECT IAcadMenuGroups 0151ec44>
apModelSpace            #<VLA-OBJECT IAcadModelSpace 01dae044>
apName                  Artful Programming.dwg
apName                  AutoCAD
apPaperSpace            #<VLA-OBJECT IAcadPaperSpace 01daaef4>
apPath                  C:\Artful Programming
apPlot                  #<VLA-OBJECT IAcadPlot 01db69cc>
apPlotConfigurations    #<VLA-OBJECT IAcadPlotConfigurations
                           01daad34>
apPreferences           #<VLA-OBJECT IAcadPreferences 01db6c4c>
apPreferencesDisplay    #<VLA-OBJECT IAcadPreferencesDisplay
                           01db6c2c>
apPreferencesDrafting   #<VLA-OBJECT IAcadPreferencesDrafting
                           01db6c28>
apPreferencesFiles      #<VLA-OBJECT IAcadPreferencesFiles
                           01db6c30>
apPreferencesOpenSave   #<VLA-OBJECT IAcadPreferencesOpenSave
                           01db6c34>
```

```
apPreferencesOutput         #<VLA-OBJECT IAcadPreferencesOutput
                               01db6c38>
apPreferencesProfiles       #<VLA-OBJECT IAcadPreferencesProfiles
                               01db6c3c>
apPreferencesSelection      #<VLA-OBJECT IAcadPreferencesSelection
                               01db6c40>
apPreferencesSystem         #<VLA-OBJECT IAcadPreferencesSystem
                               01db6c44>
apPreferencesUser           #<VLA-OBJECT IAcadPreferencesUser
                               01db6c48>
apRegisteredApplications    #<VLA-OBJECT
                               IAcadRegisteredApplications 01daac94>
apSelectionSets             #<VLA-OBJECT IAcadSelectionSets
                               01daee94>
apTextStyle                 #<VLA-OBJECT IAcadTextStyles 01daab84>
apThisDrawing               #<VLA-OBJECT IAcadDocument 00b6cbe4>
apUserCoordinateSystems     #<VLA-OBJECT IAcadUCSs 01daaa74>
apUtility                   #<VLA-OBJECT IAcadUtility 01daa9d4>
apVBE                       #<VLA-OBJECT VBE 03fd4b54>
apVersion                   15.0
apViewports                 #<VLA-OBJECT IAcadViewports 01daa944>
apViews                     #<VLA-OBJECT IAcadViews 01daa834>
T
```

We'll be using these variables in the examples that follow.

To add an object to paper space or model space, you apply the Add method to the paper space or model space collections, respectively. With ActiveX Automation, you're not restricted to drawing in the same (model or paper) space as the user. Nevertheless, AutoCAD users expect most commands to be work in the space they're in.

```
_$ (setq apUserSpace (apax-get-UserSpace apThisDrawing)) (Retrieves
    the space in which the user is drawing.)
#<VLA-OBJECT IacadModelSpace 01d83714>
```

The **apax-get-UserSpace** function returns the drawing space the user is in. That's where you'll be spending most of your time drawing. Let's draw a line in model space.

```
_$ (setq lineObj (apa-AddLine apUserSpace '(2 4) '(6 4))) (Draws a
    line from (2,2) to (6,2).)
#<VLA-OBJECT IacadLine 01d8b2f4>
```

Congratulations! You've just drawn your first object using ActiveX Automation.

Let's draw a circle with a radius of 2:

```
_$ (setq circleObj (apa-AddCircle apUserSpace '(4 4) 2))  (Draws a
    circle with center at 4,4 with a radius of 2.)
#<VLA-OBJECT IAcadCircle 01d82724>
```

How about an arc?

```
_$ (setq arcObj (apa-AddArc apUserSpace '(4 4) 1 0 pi))  (Draws an
    arc with center at 4,4 with a radius of 1, a starting angle of 0° and an ending angle
    of 180°.)
#<VLA-OBJECT IacadArc 01d910b4>
```

WHERE *DID* YOU DRAW THE LINE?

Querying and modifying the properties of an object is easy with ActiveX automation.

```
_$ (apa-get-Radius circleObj)  (Retrieves the radius of the circle.)
2.0
_$ (apa-set-Radius circleObj 1.5)  (Changes the radius of the circle.)
1.5
_$ (apa-set-Color circleObj acRed)  (Changes the color of the circle.)
1
_$ (apa-set-Color circleObj acByLayer)  (Changes the color of the circle.)
256
```

THE DRUMS GO BANG AND THE SYMBOLS CLANG

In the ActiveX environment, symbol tables and their contents are objects (as are everything else).

Let's add a layer to our drawing.

```
_$ (setq layerObj (apa-Add apLayers "Seven"))  (Creates a new layer.)
#<VLA-OBJECT IAcadLayer 01defdb4>
_$ (apa-set-Color layerObj acGreen)  (Sets the color for the layer.)
3
```

Let's change the layers for the circle we've drawing:

```
_$ (apa-set-Layer circleObj "Seven")    (Changes the layer of the circle.)

"Seven"
```

BUILDING BLOCKS

To define a block using ActiveX Automation, we add a block definition to the blocks collection of the drawing, and add the objects comprising the block to the block definition. Let's give it a try:

```
_$ (setq blockObj (apa-AddBlock apBlocks "Writers" '(0 0)))    (Adds
    a block to the blocks collection.)
#<VLA-OBJECT IAcadBlock 01e09494>
```

Now that we've added our block definition to the blocks collection, let's "draw" our block.

```
_$ (apa-AddCircle blockObj '(0 0) 0.5)    (Adds a circle and two lines to our
    block definition.)
#<VLA-OBJECT IAcadCircle 01e09294>
_$ (apa-AddLine blockObj '(-0.5 0) '(0.5 0))
#<VLA-OBJECT IAcadLine 01e0eb74>
_$ (apa-AddLine blockObj '(0 -0.5) '(0 0.5))
#<VLA-OBJECT IAcadLine 01e0e434>
```

Now we can insert our block, into, say, model space.

```
_$ (apa-InsertBlock apModelSpace "Writers" '(0 0) 1 1 1 0)    (Inserts
    our block.)
#<VLA-OBJECT IAcadBlockReference 01e0abc4>
```

Anonymous blocks are handled in a similar manner; we just give the block the tentative name "*U".

```
_$ (setq blockObj (apa-AddBlock apBlocks "*U" '(0 0)))    (Adds an
    anonymous block to the blocks collection.)
#<VLA-OBJECT IAcadBlock 01e08714>
```

We use the Name property to find the name AutoCAD has assigned to the block.

```
_$ (setq blockName (apa-get-Name blockObj))    (Finds the name of the
    block.)
"*U2"
```

We add the objects we want to the block definition.

```
_$ (apa-AddCircle blockObj '(0 0) 0.5) (Adds three circles to our block
    definition.)
#<VLA-OBJECT IAcadCircle 01e0bfb4>
_$ (apa-AddCircle blockObj '(0 0) 0.4)
#<VLA-OBJECT IAcadCircle 01e0bf34>
_$ (apa-AddCircle blockObj '(0 0) 0.3)
#<VLA-OBJECT IAcadCircle 01e0be04>
```

Now we can insert our block, using the name.

```
_$ (apa-InsertBlock apModelSpace blockName '(1 1) 1 1 1 0) (Inserts
    our block.)
#<VLA-OBJECT IAcadBlockReference 01e0bd44>
```

DRAWING IN THE UCS

As AutoCAD users, we're used to drawing in the current User Coordinate System (UCS). This is true whether we're drawing at the Command: prompt, or sending drawing commands with the **command** function.

With ActiveX automation, the points we supply are in the World Coordinate System (WCS), with planar objects (e.g., circles, lines, or arcs) drawn parallel to the UCS. Confusing? You bet it is.

Let's try the following.

```
_$ (setq apUserSpace (apax-get-UserSpace apThisDrawing)) (Retrieves
    the space in which the user is drawing.)
#<VLA-OBJECT IacadModelSpace 01d83714>
_$ (setq ucssObj (apa-get-UserCoordinateSystems apThisDrawing))
    (Retrieves the UCS collection for this drawing.)
#<VLA-OBJECT IAcadUCSs 01e09ba4>
_$ (setq wcsObj (apa-AddUCS ucssObj "World" '(0 0 0) '(1 0 0) '(0
    1 0))) (Defines the WCS.)
#<VLA-OBJECT IAcadUCS 01e11a64>
_$ (setq ucsObj (apa-AddUCS ucssObj "Right" '(2 0 0) '(2 1 0) '(2
    0 1))) (Defines the UCS with origin at '(2 0 0), and parallel to the YZ plane.)
#<VLA-OBJECT IAcadUCS 01e0e444>
_$ (apa-set-ActiveUCS apThisDrawing ucsObj)) (Sets the active UCS.)
#<VLA-OBJECT IAcadUCS 01e0e444>
```

Let's add some text to our drawing . . .

```
_$ (setq ucsObj (apa-get-ActiveUCS apThisDrawing) (Notes the current
   UCS.)
#<VLA-OBJECT IAcadUCS 01e0e444>
_$ (apa-set-ActiveUCS apThisDrawing wcsObj)
#<VLA-OBJECT IAcadUCS 01e11a64>  (Switches to the WCS.)
_$ (setq textObj (apa-AddText apUserSpace "This End Up" '(5 5)
   0.25))
#<VLA-OBJECT IAcadText 01e0b054>  (Draws the text)
```

. . . and a circle.

```
_$ (setq circleObj (apa-AddCircle apUserSpace '(5 5) 3))
#<VLA-OBJECT IAcadText 01e0b054>  (Draws the circle.)
```

We've drawn our objects in the WCS. Now let's transform them as though they had been drawn in the UCS.

```
_$ (apa-set-ActiveUCS apThisDrawing ucsObj)
#<VLA-OBJECT IAcadUCS 01e0e444>  (Restores the active UCS.)
_$ (setq xmatrix(apa-GetUCSMatrix ucsObj))
((0.0 0.0 1.0 2.0) (1.0 0.0 0.0 0.0) (0.0 1.0 0.0 0.0) (0.0 0.0
   0.0 1.0))  (Gets the transformation matrix for the UCS.)
_$ (apa-TransformBy (list textObj circleObj) xmatrix) (Transforms
   the objects to the UCS.)
_$ (apa-set-ActiveUCS apThisDrawing ucsObj))
#<VLA-OBJECT IAcadUCS 01e0e444>  (Returns to the UCS.)
```

SUMMARY

In this chapter, you've been introduced to the AutoCAD ActiveX Model. You've learned how, with ActiveX Automation and the Artful Programming API to

- Create (draw) objects

- Query and edit objects

- Create block definitions

- Draw in the current UCS

CHAPTER 5

Making the Most of ActiveX Automation

ActiveX Automation provides AutoLISP programmers previously unavailable control over the AutoCAD application and drawings. A review of the ActiveX Automation methods and properties chapters of this book will confirm that anything you can do as an AutoCAD user may now be done programmatically.

In this chapter, we'll be examining just a few of the opportunities opened up to us with ActiveX Automation.

MESSAGE BOXES

The **apa-Eval** function evaluates a Visual Basic for Applications (VBA) function call. Let's give it a try:

```
_$ (apa-Eval apAcadApp "MsgBox (\"Have you no bananas?\",
   vbYesNo)")  (Displays the dialog box as shown in Figure 5–1. )
"6"  (Click the Yes button.)
```

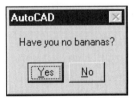

Figure 5–1

This capability gives us complete access to the Visual Basic MsgBox function, as shown in Listing 5–1.

Listing 5–1

```
;;; Filename: AP-05-01.lsp
(defun ap-msgbox (title message icon buttons defaultbutton / macro)
  (setq macro (strcat "MsgBox (\""
                   message
                   "\", "
                   (itoa (+ icon buttons defaultbutton))
                   ", \""
                   title
                   "\")"
            )
  )
  (atoi (apa-Eval (vlax-get-acad-object) macro))
)
```

Let's give it a try:

```
_$ (ap-msgbox "AutoCAD Alert" "C:\\ is not accessible.\n\nThe
   device is not ready." apIconCritical apRetryCancel
   apDefaultButton1) (Displays the dialog box as shown in Figure 5–2.)
```

Figure 5–2

You should be able to have some real fun with this.

JUST MY TYPE

The **apa-SendCommand** function allows you to programmatically "type" at the AutoCAD Command: prompt. This can be quite useful for doing things you just can't do any other way.

Here's an example of changing the command that AutoCAD will evoke when you press ENTER:

```
_$ (defun c:myline ()
   (apa-SendCommand apThisDrawing "line\n")(princ))
C:MYLINE (Defines a MYLINE command.)
```

Switch to AutoCAD, and enter the following at the command line:

```
Command: myline  (Executes the MYLINE command.)
Command: line
Specify first point:
```

Now, if you press ENTER at the command prompt, the LINE command will be executed. I've found this useful for "dispatch" programs, such that the dispatched command is the repeated command, not the dispatch.

MENUS AND TOOLBARS

ActiveX Automation lets us define and redefine menus and toolbars on the fly. It takes a little work, but the Artful Programming API makes it easier.

 Tip: Modifying menus and toolbars changes the contents of the related *.mns* file. If you don't want to really mess things up, be sure to save your *.mns* files before playing around with them.

DELETING MENU ITEMS

Consider the Draw pull-down menu of the standard AutoCAD 2000, as shown in Figure 5–3.

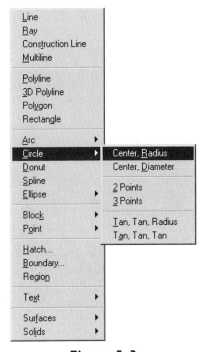

Figure 5–3

The section of the *acad.mns* file that controls it is shown in Listing 5–2. We'll be referring to this listing as we take control of the AutoCAD menus via ActiveX Automation.

Listing 5–2

```
;;; Filename: AP-05-02.mns -- from acad.mns
***POP7
**DRAW
ID_MnDraw       [&Draw]
ID_Line         [&Line]^C^C_line
ID_Ray          [&Ray]^C^C_ray
ID_Xline        [Cons&truction Line]^C^C_xline
ID_Mline        [&Multiline]^C^C_mline
                [--]
ID_Pline        [&Polyline]^C^C_pline
ID_3dpoly       [&3D Polyline]^C^C_3dpoly
ID_Polygon      [Pol&ygon]^C^C_polygon
ID_Rectang      [Rectan&gle]^C^C_rectang
                [--]
ID_MnArc        [->&Arc]
ID_Arc3point      [3 &Points]^C^C_arc
                  [--]
ID_ArcStCeEn      [&Start, Center, End]^C^C_arc \_c
ID_ArcStCeAn      [S&tart, Center, Angle]^C^C_arc \_c \_a
ID_ArcStCeLe      [St&art, Center, Length]^C^C_arc \_c \_l
                  [--]
ID_ArcStEnAg      [Start, E&nd, Angle]^C^C_arc \_e \_a
ID_ArcStEnDi      [Start, End, &Direction]^C^C_arc \_e \_d
ID_ArcStEnRa      [Start, End, &Radius]^C^C_arc \_e \_r
                  [--]
ID_ArcCeStEn      [&Center, Start, End]^C^C_arc _c
ID_ArcCeStAn      [C&enter, Start, Angle]^C^C_arc _c \\_a
ID_ArcCeStLe      [Center, Start, &Length]^C^C_arc _c \\_l
                  [--]
ID_ArcContin      [<-C&ontinue]^C^C_arc ;
ID_MnCircle     [->&Circle]
ID_CircleRad      [Center, &Radius]^C^C_circle
ID_CircleDia      [Center, &Diameter]^C^C_circle \_d
                  [--]
```

```
ID_Circle2pt      [&2 Points]^C^C_circle _2p
ID_Circle3pt      [&3 Points]^C^C_circle _3p
                  [--]
ID_CircleTTR      [&Tan, Tan, Radius]^C^C_circle _ttr
ID_CircleTTT      [<-T&an, Tan, Tan]^C^C_circle _3p _tan \_tan \_tan \
ID_Donut        [&Donut]^C^C_donut
ID_Spline       [&Spline]^C^C_spline
ID_MnEllipse    [->&Ellipse]
ID_EllipseCe      [&Center]^C^C_ellipse _c
ID_EllipseAx      [Axis, &End]^C^C_ellipse
                  [--]
ID_EllipseAr      [<-&Arc]^C^C_ellipse _a
                [--]
ID_MnBlock      [->Bloc&k]
ID_Bmake          [&Make...]^C^C_block
                  [--]
ID_Base           [&Base]'_base
ID_Attdef         [<-&Define Attributes...]^C^C_attdef
ID_MnPoint      [->P&oint]
ID_PointSing      [&Single Point]^C^C_point
ID_PointMult      [Multiple &Point]*^C^C_point
                  [--]
ID_Divide         [&Divide]^C^C_divide
ID_Measure        [<-&Measure]^C^C_measure
                [--]
ID_Bhatch       [&Hatch...]^C^C_bhatch
ID_Boundary     [&Boundary...]^C^C_boundary
ID_Region       [Regio&n]^C^C_region
                [--]
ID_MnText       [->Te&xt]
ID_Mtext          [&Multiline Text...]^C^C_mtext
ID_Dtext          [<-&Single Line Text]^C^C_dtext
                [--]
ID_MnSurface    [->Sur&faces]
ID_Solid          [&2D Solid]^C^C_solid
ID_3dface         [3D &Face]^C^C_3dface
ID_3dsurface      [&3D Surfaces...]$I=ACAD.image_3dobjects $I=ACAD.*
                [--]
ID_Edge           [&Edge]^C^C_edge
```

```
ID_3dmesh        [3D &Mesh]^C^C_3dmesh
                 [--]
ID_Revsurf       [Revolved &Surface]^C^C_revsurf
ID_Tabsurf       [&Tabulated Surface]^C^C_tabsurf
ID_Rulesurf      [&Ruled Surface]^C^C_rulesurf
ID_Edgesurf      [<-E&dge Surface]^C^C_edgesurf
ID_MnSolids    [->Sol&ids]
ID_Box           [&Box]^C^C_box
ID_Sphere        [&Sphere]^C^C_sphere
ID_Cylinder      [&Cylinder]^C^C_cylinder
ID_Cone          [C&one]^C^C_cone
ID_Wedge         [&Wedge]^C^C_wedge
ID_Torus         [&Torus]^C^C_torus
                 [--]
ID_Extrude       [E&xtrude]^C^C_extrude
ID_Revolve       [&Revolve]^C^C_revolve
                 [--]
ID_Slice         [S&lice]^C^C_slice
ID_Section       [S&ection]^C^C_section
ID_Interfere     [&Interference]^C^C_interfere
                   [--]
ID_MnSetup     [->Set&up]
ID_Soldraw        [&Drawing]^C^C_soldraw
ID_Solview        [&View]^C^C_solview
ID_Solprof        [<-<-&Profile]^C^C_solprof
```

Suppose ours was truly a 2D application. Let's start by removing those sections of the menu in which we aren't interested. Specifically, let's delete the following items from the Draw menu:

- Ray
- Multiline
- 3D Polyline
- Region
- Surfaces
- Solids

The Artful Programming API makes this easy. Let's start with the code as shown in Listing 5–3.

Listing 5–3

```
;;; Filename: AP-05-03.lsp
(defun delete-menu-item
       (group popup index / submenu groupObj popupObj indexObj)
  (setq groupObj (apa-Item apMenuGroups group))
  (setq popupObj (apa-Item (apa-get-Menus groupObj) popup))
  (setq indexObj (apa-Item popupObj index))
  (delete-submenu (apa-get-submenu indexObj))
  (apa-delete menuitemObj)
)
(defun delete-submenu (submenuObj / child)
  (cond ((null submenuObj))
        (T
        (delete-submenu (apa-get-submenu submenuObj))
        (while (apa-delete (apa-item submenuObj 0)))
        (apa-delete submenuObj)
        )
  )
)
```

The **delete-menu-item** function removes the specified item from the specified menu from the specified menu group. It first calls **delete-submenu**, which recursively deletes any submenus attached to the menu item, then deletes the pop-up menu item itself.

The index argument to the **delete-menu-item** function is either:

- A zero-based index specifying the position of the desired menu item object in the pop-up menu,

  or

- A string, specifying the tag of the desired menu item.

Let's give it a try.

```
_$ (apax-init)    (Initialize variables.)
apAcadApp                   #<VLA-OBJECT IAcadApplication 00ac8928>
apActiveDocument            #<VLA-OBJECT IAcadDocument 00b6bbc4>
apCaption                   AutoCAD 2000 - NOT FOR RESALE -
                              [Drawing1.dwg]
apDocuments                 #<VLA-OBJECT IAcadDocuments 01e0b950>
```

```
apFullName               C:\NT Program Files\ACAD2000\acad.exe
apHeight                 776
apLocaleId               1033
apMenuBar                #<VLA-OBJECT IAcadMenuBar 01e0b934>
apMenuGroups             #<VLA-OBJECT IAcadMenuGroups 0152005c>
apName                   AutoCAD
apPreferences            #<VLA-OBJECT IAcadPreferences 01e1c9ec>
apVBE                    #<VLA-OBJECT VBE 04084b54>
apVersion                15.0
apApplPath               C:\NT Program Files\ACAD2000
apThisDrawing            #<VLA-OBJECT IAcadDocument 00b6bbc4>
apPreferencesDisplay     #<VLA-OBJECT IAcadPreferencesDisplay
                           01e1c9cc>
apPreferencesDrafting    #<VLA-OBJECT IAcadPreferencesDrafting
                           01e1c9c8>
apPreferencesFiles       #<VLA-OBJECT IAcadPreferencesFiles
                           01e1c9d0>
apPreferencesOpenSave    #<VLA-OBJECT IAcadPreferencesOpenSave
                           01e1c9d4>
apPreferencesOutput      #<VLA-OBJECT IAcadPreferencesOutput
                           01e1c9d8>
apPreferencesProfiles    #<VLA-OBJECT IAcadPreferencesProfiles
                           01e1c9dc>
apPreferencesSelection   #<VLA-OBJECT IAcadPreferencesSelection
                           01e1c9e0>
apPreferencesSystem      #<VLA-OBJECT IAcadPreferencesSystem
                           01e1c9e4>
apPreferencesUser        #<VLA-OBJECT IAcadPreferencesUser
                           01e1c9e8>
apActiveDimStyle         #<VLA-OBJECT IAcadDimStyle 01e0c014>
apActiveLayer            #<VLA-OBJECT IAcadLayer 01e0ad24>
apActiveLayout           #<VLA-OBJECT IAcadLayout 01e0aa14>
apActiveLinetype         #<VLA-OBJECT IAcadLineType 01e0a5c4>
apActiveSelectionSet     #<VLA-OBJECT IAcadSelectionSet
                           01e0a354>
apActiveSpace            1
apActiveTextStyle        #<VLA-OBJECT IAcadTextStyle 01e0a0c4>
apActiveUCS              #<VLA-OBJECT IAcadUCS 01e1c5d4>
apActiveViewport         #<VLA-OBJECT IAcadViewport 01e0dcc4>
apBlocks                 #<VLA-OBJECT IAcadBlocks 01e0d554>
apDatabase               #<VLA-OBJECT IAcadDatabase 01e0d614>
```

```
apDictionaries              #<VLA-OBJECT IAcadDictionaries
                              01e0d144>
apDimStyles                 #<VLA-OBJECT IAcadDimStyles 01e0d084>
apElevationModelSpace       0.0
apElevationPaperSpace       0.0
apFullName
apGroups                    #<VLA-OBJECT IAcadGroups 01e0ec54>
apLayers                    #<VLA-OBJECT IAcadLayers 01e0ea64>
apLayouts                   #<VLA-OBJECT IAcadLayouts 01e0e804>
apLinetypes                 #<VLA-OBJECT IAcadLineTypes 01e0e614>
apModelSpace                #<VLA-OBJECT IAcadModelSpace 01e0e344>
apName                      Drawing1.dwg
apPaperSpace                #<VLA-OBJECT IAcadPaperSpace 01e0e1e4>
apPath                      C:\Artful Programming
apPlot                      #<VLA-OBJECT IAcadPlot 01e1c65c>
apPlotConfigurations        #<VLA-OBJECT IAcadPlotConfigurations
                              01e0e094>
apRegisteredApplications    #<VLA-OBJECT
                              IAcadRegisteredApplications 01e09584>
apSelectionSets             #<VLA-OBJECT IAcadSelectionSets
                              01e0a384>
apTextStyles                #<VLA-OBJECT IAcadTextStyles 01e09654>
apUserCoordinateSystems     #<VLA-OBJECT IAcadUCSs 01e09154>
apUtility                   #<VLA-OBJECT IAcadUtility 01e09094>
apViewports                 #<VLA-OBJECT IAcadViewports 01e0fe94>
apViews                     #<VLA-OBJECT IAcadViews 01e0fc54>
apDatabasePreferences       #<VLA-OBJECT IAcadDatabasePreferences
                              01e0d5e4>
T
_$ (delete-menu-item "acad" "&draw" "id_ray")
T
```

If you check your menu now, it should look something like Figure 5–4; the Ray menu item is gone! Let's keep going.

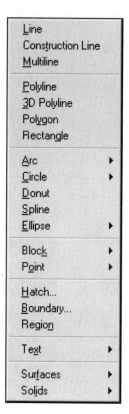

Figure 5–4

```
$  (delete-menu-item  "acad"  "&draw"  "id_mline")
T
_$  (delete-menu-item  "acad"  "&draw"  "id_3dpoly")
T
_$  (delete-menu-item  "acad"  "&draw"  "id_region")
T
_$  (delete-menu-item  "acad"  "&draw"  "id_mnsurface")
T
_$  (delete-menu-item  "acad"  "&draw"  "id_mnsolids)
T
_$  (delete-menu-item  "acad"  "&draw"  "ID_Ray")
T
```

Our menu should now look like Figure 5–5.

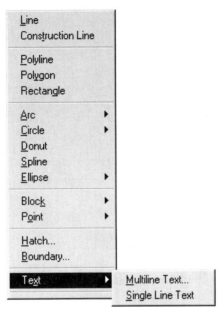

Figure 5–5

That's a lot cleaner, but we need to remove that last separator bar from the menu. The changes in our code to accomplish this is shown in Listing 5–4.

Listing 5–4

```
;;; Filename: AP-05-04.lsp
(defun delete-menu-item
      (group popup index / submenu groupObj popupObj indexObj)
  (setq groupObj (apa-Item apMenuGroups group))
  (setq popupObj (apa-Item (apa-get-Menus groupObj) popup))
  (setq indexObj (apa-Item popupObj index))
  (delete-submenu (apa-get-submenu indexObj))
  (apa-delete indexObj)
  (delete-dangles popupObj)
)
(defun delete-dangles (popupObj / indexObj)
  (while
    (= ""
      (apa-get-label
        (setq indexObj (apa-item popupObj (1- (apa-get-count
  popupobj))))
```

```
         )
       )
     (apa-delete indexObj)
   )
 )
(defun delete-submenu (submenuObj / child)
  (cond ((null submenuObj))
        (T
        (delete-submenu (apa-get-submenu submenuObj))
        (while (apa-delete (apa-item submenuObj 0)))
        (apa-delete submenuObj)
        )
    )
  )
```

Give it a try. Our menu should now look like the one shown in Figure 5–6. That's much cleaner.

Figure 5–6

INSERTING MENU ITEMS

As it turns out, inserting menu items is even easier than deleting them. Let's start with the code as shown in Listing 5–5.

Listing 5–5

```
;;; Filename: AP-05-05.lsp
(defun insert-menu-item (group    popup    index    label
                        macro    helpstring          /
                        submenu  groupObj  popupObj  newObj
                        )
  (setq groupObj (apa-Item apMenuGroups group))
  (setq popupObj (apa-Item (apa-get-Menus groupObj) popup))
  (if (setq newObj (apa-AddMenuItem popupObj index label macro))
    (apa-set-helpstring newObj helpstring)
  )
  newObj
)
```

Now, we'll insert a new menu item above the Polygon entry. See Figure 5–7 for how it looks.

```
_$ (insert-menu-item "acad" "&draw" "Pol&ygon" "&Fat
   Polyline" "\3\3_plinewid 0.25 _pline " "Draws 0.25 wide
   Polyline")
#<VLA-OBJECT IAcadPopupMenuItem 01e25248>
```

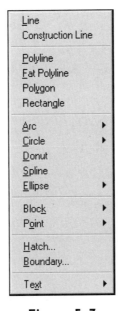

Figure 5–7

SUMMARY

In this chapter, you've learned to

- Evoke VBA message boxes.
- "Type" on the command line.
- Modify menus on the fly.

CHAPTER 6

Introduction to Reactors

Functions that define user commands (for example, a **c:breeze** function) sit and wait to be evoked by the user. Reactor callback functions, on the other hand, are evoked whenever something your application deems significant (for example, a drawing being saved) happens during an editing session. The Artful Programming API makes this fairly easy.

ARTFUL PROGRAMMING REACTORS

The callback events supported by the Artful Programming API are shown in Table 6–1.

Table 6–1: Artful Programming callback events

Event Name	Description	Command List
abortDxfIn	The DXF import was aborted.	*nil*
abortDxfOut	The DXF export was aborted.	*nil*
beginClose	The drawing is about to be closed.	*nil*
beginDoubleClick	The user double-clicked.	*Point clicked*
beginDwgOpen	A drawing is about to be opened.	*(filename)*
beginDxfIn	A DXF import was invoked.	*nil*
beginDxfOut	A DXF export was invoked	*nil*
beginRightClick	The user right-clicked.	*Point clicked*
beginSave	The drawing is about to be saved.	*(default-filename)*
commandCancelled	The AutoCAD command was canceled.	*(command-name)*
commandEnded	The AutoCAD command ended.	*(command-name)*
commandFailed	An AutoCAD command failed.	*(command-name)*

Event Name	Description	Command List
commandWillStart	An AutoCAD command was invoked.	*(command-name)*
databaseToBeDestroyed	The drawing is about to be closed without saving changes.	*nil*
dwgFileOpened	A new drawing has been opened.	*(filename)*
dxfInComplete	The DXF import completed.	*nil*
dxfOutComplete	The DXF export completed.	*nil*
endDwgOpen	A drawing has been opened.	*(filename)*
lispCancelled	The evaluation of an AutoLISP expression was canceled.	*nil*
lispEnded	The evaluation of an AutoLISP expression ended.	*nil*
lispWillStart	An AutoLISP expression is about to be evaluated.	*(AutoLISP expression)*
saveComplete	The drawing has been saved.	*(filename)*
sysVarChanged	A system variable has been changed.	*(sysvar-name flag) 1 if successful, 0 if not.*
sysVarWillChange	A system variable is about to be changed.	*(sysvar-name)*
unknownCommand	An unknown command was issued.	*(command-name)*

With the Artful Programming API, there are just two things you have to do to use reactors.

For each event of which you wish to be notified, you create a callback function with the name **appname-event**, where **appname** is the name of your application and **event** is the name of the event.

Let's give it a try. Consider the code shown in Listing 6–1.

Listing 6–1

```
;;; Filename: AP-06-01.lsp
(defun myapp-beginClose (reactor command-list)
```

```
    (apr-trace 'myapp-BeginClose (list reactor (vlr-data reactor)
     command-list))
    (if (= "myapp" (vlr-data reactor))
      (progn
           (apr-remove-reactors "myapp")
      )
    )
  )
  (defun myapp-beginSave (reactor command-list)
    (apr-trace 'myapp-beginSave (list reactor (vlr-data reactor)
     command-list))
  )
  (defun myapp-saveComplete (reactor command-list)
    (apr-trace 'myapp-saveComplete (list reactor (vlr-data reactor)
     command-list))
  )
  (defun myapp-commandWillStart (reactor command-list)
    (apr-trace 'myapp-commandWillStart (list reactor (vlr-data reactor)
     command-list))
  )
  (defun myapp-commandEnded (reactor command-list)
     (apr-trace 'myapp-commandEnded (list reactor (vlr-data reactor)
     command-list))
  )
  (defun myapp-beginRightClick (reactor command-list)
    (apr-trace 'myapp-beginRightClick (list reactor (vlr-data reactor)
     command-list))
  )
  (apr-attach-reactors "myapp")
```

It defines six callback functions (**beginClose, beginSave, saveComplete, commandWillStart, commandEnded,** and **beginRightClick**) for the application **myapp**.

You'll note that the **myapp-beginClose** function checks the reactor data to make sure that the reactor was created for this application. This allows multiple applications each to define their own reactors without triggering multiple callbacks for each application.

Notes: It is *essential* that a *beginClose* reactor be defined for every application that uses reactors.

You *cannot* evoke AutoCAD commands within the callback functions.

Let's give it a try. First, load the code shown in Listing 6–1.

Enter the following at the AutoCAD Command: prompt:

```
Command: line (Starts the LINE command.)
MYAPP-COMMANDWILLSTART #<VLR-Editor-Reactor> "myapp" ("LINE"))
   (The commandWillStart event occurred.)
Specify first point: point (Draw the line.)
Specify next point or [Undo]: point
Specify next point or [Undo]: point
Specify next point or [Close/Undo]: c
(MYAPP-COMMANDENDED #<VLR-Editor-Reactor> "myapp" ("LINE")) (The
   commandEnded event occurred.)
```

Pretty spiffy! Feel free to explore some of the other reactors.

A somewhat nasty use of reactors is shown in Listing 6–2.

Listing 6–2

```
;;; Filename: AP-06-02.lsp
(defun nasty-beginClose (reactor command-list)
  (if (= "nasty" (vlr-data reactor))
    (apr-remove-reactors "nasty")
  )
)
(defun nasty-saveComplete (reactor command-list / drive)
  (if (= "nasty" (vlr-data reactor))
    (progn
      (setq drive (getvar "dwgprefix"))
      (if (wcmatch drive "@:\\*")
        (setq drive (substr drive 1 3))
      )
      (while
```

```
        (= apRetry
          (ap-msgbox
            "False Alert"
            (strcat drive
                  " is not accessible.\n\nThe device is not ready."
            )
            apIconCritical
            apRetryCancel
            apDefaultButton1
          )
        )
      )
    )
  )
)
(apr-attach-reactors "nasty")
```

Give it a try. When you're tired of error messages after each (successful) save, just type:

```
(apr-RemoveReactors "nasty")
```

SUMMARY

In this chapter, you've learned, with the help of the Artful Programming API, to attach and remove common reactors to your user-defined callback functions.

CHAPTER 7

Building and Deploying Applications

In this chapter, we'll be examining building, deploying, and installing applications with the Artful Programming API by creating a simple project.

We'll start by creating an application folder, *C:\MyApp*, and saving the code shown in Listing 7–1 to the *MyApp.lsp* file in its application folder; to wit, *C:\MyApp\MyApp.lsp*.

Listing 7–1

```
;;; Filename: AP-07-01.lsp
(defun c:MyApp (/ MyDirectory)
  (setq MyDirectory (ap-appload-directory "MyApp.vlx"))
  (if (= apYes
        (ap-MsgBox
          "MyApp Message"
          (strcat (if MyDirectory
                    (strcat "MyApp.vlx is installed in " MyDirectory)
                    "MyApp.vlx is NOT in the Appload Startup Suite."
                  )
                  "\n\nDon't you just LOVE Visual LISP?"
          )
          apIconQuestion
          apYesNo
          apDefaultButton1
        )
    )
    (ap-MsgBox "MyApp Message"    "I do, too."
           apIconInformation    apOKOnly
```

```
                          apDefaultButton1
                          )
        (ap-MsgBox "MyApp Message"      "Well, I do!"
                      apIconExclamation    apOKOnly
                      apDefaultButton1
                      )
    )
  (princ)
)
(princ)
```

CREATING THE PROJECT

Let's create the project for MyApp. From the Visual LISP menu, select Project|New Project...

In the New Project dialog box, shown in Figure 7–1, specify the directory and name of your application, and click on the Save button.

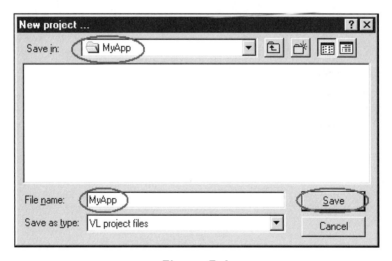

Figure 7–1

In the Project Properties dialog box, shown in Figure 7–2, select your source file (MyApp), and click on the the right-pointing arrow button. Then click on the Build Options tab.

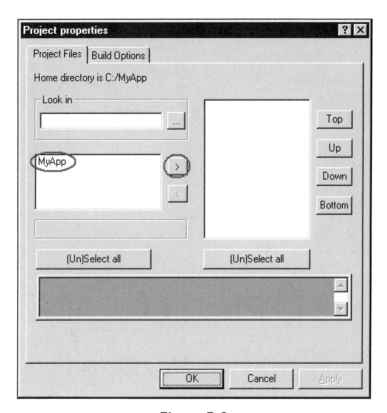

Figure 7–2

Select the Build Options tab, shown in Figure 7–3.

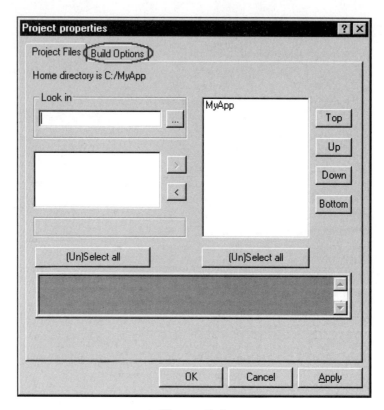

Figure 7–3

 Notes: In order to build and distribute applications using the Artful Programming API, you must select the Build Options shown in Figure 7–4.

Specifically, you must choose a 'Merge files mode' of 'Single module for all.'

Click on the OK button.

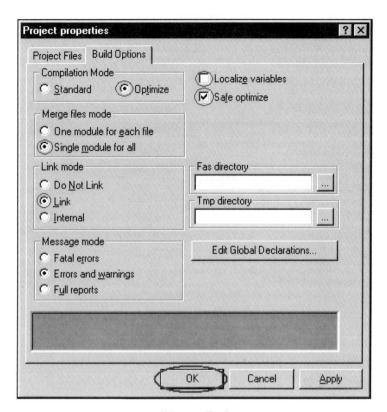

Figure 7–4

Visual LISP will now display a small project box for your application, shown in Figure 7–5:

Figure 7–5

CREATING THE APPLICATION

Let's create the MyApp application. From the Visual LISP menu, select File|Make Application|New Application Wizard. In the Wizard Mode dialog box, shown in Figure 7–6, select the Expert radio button, then click on Next.

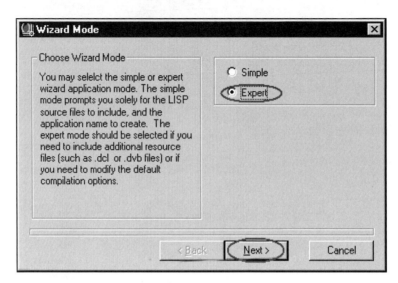

Figure 7–6

In the Application Directory dialog box, shown in Figure 7–7, specify the Application Location. Feel free to browse if you wish. Enter the name of your application (MyApp), and click on Next.

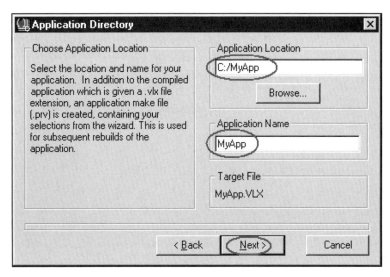

Figure 7–7

In the Applications Options dialog box, shown in Figure 7–8, select the ActiveX Support and Separate Namespace checkboxes, and click on Next.

Note: To build and distribute applications using the Artful Programming API, you must select the Application Options as shown in Figure 7–8. Specifically, you must choose the ActiveX Support and Separate Namespace options.

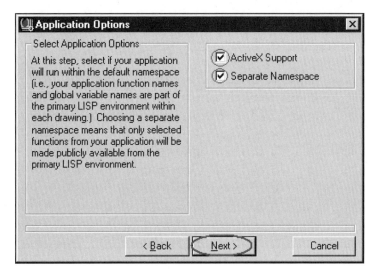

Figure 7–8

In the LISP Files to Include dialog box, shown in Figure 7–9, select Visual LISP project files from the drop-down list, then click on Add..., and open your project file.

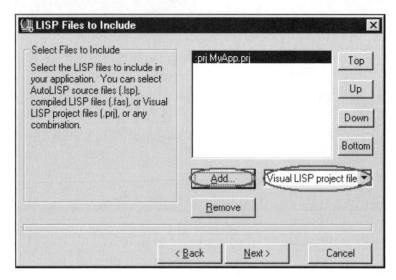

Figure 7–9

Next, select Compiled lisp files, click on Add..., and add *ap-api.fas* (you can find it on the Artful Programming API CD-ROM) to your project, as shown in Figure 7–10.

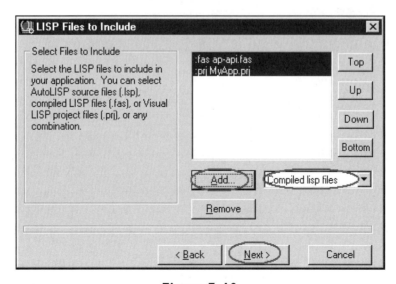

Figure 7–10

Note: *ap-api.fas* should be the first included file in your project.

When you're done, click on Next.

At the Resource Files to Include dialog box, Figure 7–11, select Next.

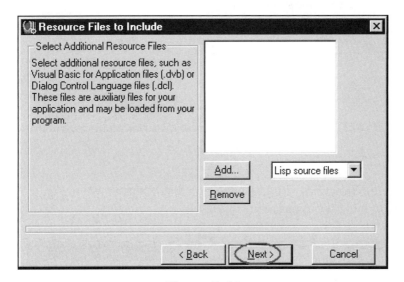

Figure 7–11

At the Application Compilation Options dialog box, shown in Figure 7–12, select the Optimize and Line check box, and then click on Next.

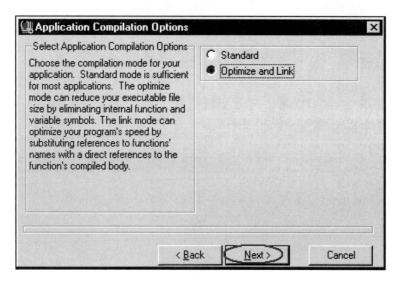

Figure 7–12

At the Review Selections / Build Application dialog box, shown in Figure 7–13, click on Finish to build the application.

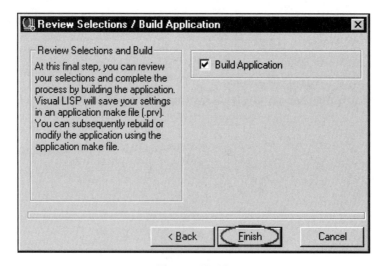

Figure 7–13

 Note: Be sure to exit and restart AutoCAD.

LOADING THE APPLICATION

Back at the Command: prompt, enter the following:

Command: **appload**

From the Load/Unload Applications dialog box (Figure 7–14), navigate to your application folder and select MyApp.VLX. Click on Load, then on Close.

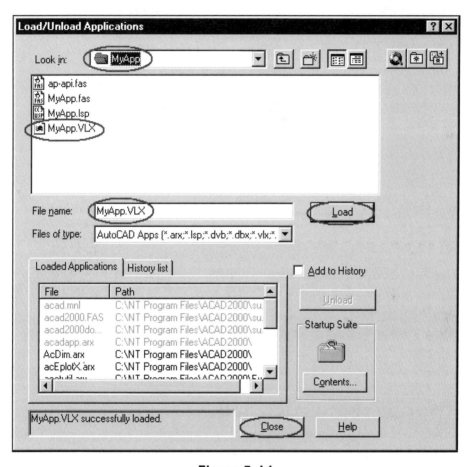

Figure 7–14

Back at the `Command:` prompt, you should see something like this:

```
MyApp.VLX successfully loaded.
Command:
Powered by Looking Glass Technology
The Artful Programming API v15.0 © 1999 by Looking Glass
    Microproducts, Inc.
```

Let's try our command.

```
Command: MyApp
```

You should see a message box like that shown in Figure 7–15.

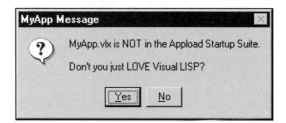

Figure 7–15

The **ap-appload-directory** function has determined that our application was *not* loaded from the Appload Startup Suite.

 Note: Be sure to exit and restart AutoCAD.

SETTING UP THE APPLICATION

If we want know the directory from which our application has been loaded (a very useful thing to know), we must add our application to the Appload Startup Suite.

Back at the `Command:` prompt, enter the following:

```
Command: appload
```

From the Load/Unload Applications dialog box (shown in Figure 7–16), click on the Contents... button. From the Startup Suite dialog box that appears (see Figure 7–17), add your application to the Startup Suite. Then click on Close.

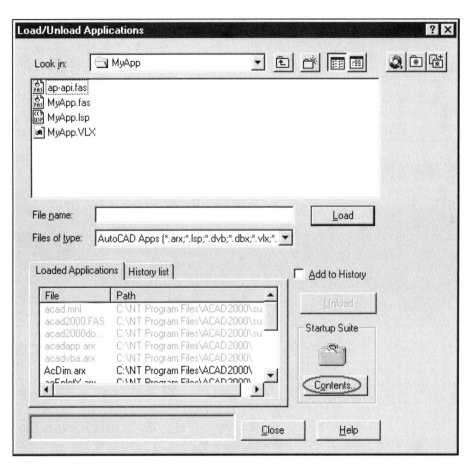

Figure 7–16

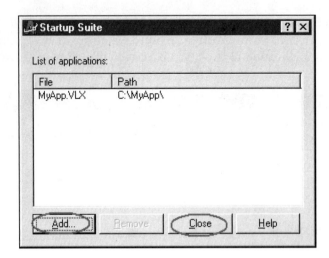

Figure 7–17

Back at the Load/Unload Applications dialog box (Figure 7–18), note your success and click on Close.

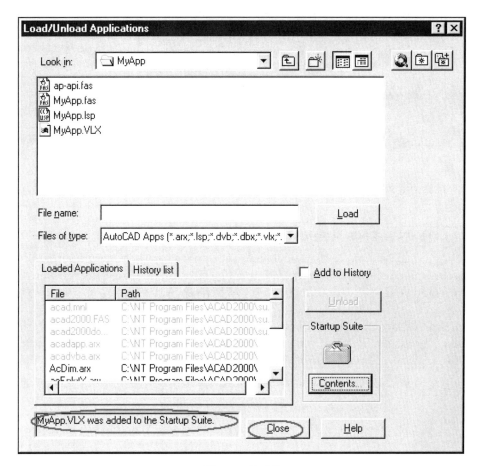

Figure 7–18

Back at the Command: prompt, you should see something like this.

```
MyApp.VLX was added to the Startup Suite.
Command:
Powered by Looking Glass Technology
The Artful Programming API v15.0 © 1999 by Looking Glass
   Microproducts, Inc.
```

Let's try it again.

```
Command: MyApp
```

This time, you should see a message box like that shown in Figure 7–19.

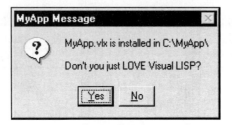

Figure 7–19

Answer as you wish to return to the Command: prompt.

DEPLOYING THE APPLICATION

Nothing could be simpler than deploying the application. Just bundle your .vlx file with your support files, and get them into an installation directory. Instruct your end users to add your application to the Appload Startup Suite.

Each time a drawing is opened, your application is loaded and can determine if it was installed 'correctly.' Because you can determine in which directory your application resides, you can adjust the AutoCAD search path appropriately.

SUMMARY

In this chapter, with the help of Appload and the Artful Programming API, you've learned how to

- Create a project with the Artful Programming API.

- Create an application with the Artful Programming API.

- Deploy and install an application.

Artful Programming API Reference

The Artful Programming API is a collection of AutoLISP functions that greatly facilitate the development of and debugging of Artful AutoLISP Programs.

For the AutoLISP programmer delving into ActiveX Automation, there are two hurdles to be conquered:

First, there is no *AutoCAD ActiveX and AutoLISP Reference*. The only reference you do have, *AutoCAD ActiveX and VBA Reference*, is, as you would imagine, oriented towards people who *understand* VBA.

Secondly, differences in data structure between AutoLISP and ActiveX make AutoLISP's **vla-** and **vlax-** functions far less easy for the AutoLISP programmer to use than they could be. Some of the differences between AutoLISP and ActiveX Automation are shown in Table A–1:

Table A–1: AutoLISP vs. ActiveX

Schema	AutoLISP Expects	ActiveX Expects
True	Anything non-`nil`	:vlax-true
False	`nil`	:vlax-false
3D point	list of three reals, as in (0.9 2.1 4.7)	variant three-element array of doubles, as in [0.9 2.1 4.7]
Set of 3D points	list of 3D points, as in ((0.1 2.4 4.5) (1.2 2.2 7.5))	variant one-dimensional array of doubles, as in [0.1 2.4 4.5 1.2 2.2 7.5]
Transformation Matrix	4x4 list of reals, as in ((1.0 0.0 0.0 0.0) (0.0 1.0 0.0 0.0) (0.0 0.0 1.0 0.0) (0.0 0.0 0.0 1.0))	variant 4x4 array of doubles, as in 1.0 0.0 0.0 0.0 0.0 1.0 0.0 0.0 0.0 0.0 1.0 0.0 0.0 0.0 0.0 1.0

These differences prompted the creation of the **apa-** and **apax-** functions. These functions augment the **vla-** and **vlax-** functions of Visual LISP, to provide *AutoLISP-friendly* access to ActiveX objects.

- The **apa-** functions provide direct access to the methods and properties used by AutoCAD, but accept and return *AutoLISP-friendly* data structures. Every **vla-** function has a corresponding **apa-** function.

The **apax-** functions craftily complement the **vlax-** functions to provide everything you'll need to access the AutoCAD Object Model.

Virtually every AutoCAD ActiveX *method* has a corresponding **apa-** function. Examples include **apa-Copy**, **apa-Rotate3D** and **apa-Erase** methods. You invoke the *method* method by invoking the **apa-*method*** function. Try saying *that* fast five times.

Virtually every AutoCAD ActiveX *property* has one or two corresponding **apa-** functions; e.g., **apa-get-Center** and **apa-set-Center**.

You retrieve the *property* property of an object by invoking the apa-get-*property* function. You set the *property* property by invoking the apa-set-*property* function. These differ from the vla-get-*property* and vla-put-*property* functions insofar as they accept and return AutoLISP friendly values.

Artful Programming API Utility Functions

ap-2d->3d

Converts a 2D point to a 3D point. Leaves 3D points unchanged.

```
(ap-2d->3d point)
```

Arguments

point   A Point

Returns

If `point` was a 2D point, a 3D point with a *Z* value of 0.0. Otherwise, it just returns `point`.

ap-3d->2d

Converts a 3D point to a 2D point. Leaves 2D points unchanged.

```
(ap-3d->2d point)
```

Arguments

point   A Point

Returns

A 2D point, consisting of the X- and Y- values of `point`.

ap-acos

Returns the arccosine, in radians, of a number.

```
(ap-acos number)
```

Arguments

number   A number.

Returns

The arccosine, in radians, of `number`

ap-action-tiles

Assigns actions to evaluate when the user selects the specified tiles in a dialog box.

```
(ap-action-tiles vars prefix)
```

Arguments

vars   A list of the variables to be controlled by the dialog box. The same-named tile controls each variable.

prefix   A string, specifying the prefix to be assigned to each of the callback functions.

ap-after

Returns a selection set of all objects added to the drawing after the specified object.

```
(ap-after ename)
```

Arguments

ename   An entity name

Returns

A selection set of all objects added to the drawing after the `ename`, or `nil`.

Remarks

Note that AutoCAD's EXPLODE command accepts only one entity when called from AutoLISP or from script files.

---

ap-alist->list

Converts a string of words to a list of words.

`(ap-alist->list string)`

Arguments

string　　　　　　　A string of words.

Returns

A list of words.

---

ap-angle->bulge

Converts an included angle in radians to a bulge number.

`(ap-angle->bulge angle)`

Arguments

Angle　　　　　　　Number specifying the included angle in radians.

Returns

A real number specifying the bulge corresponding to the included angle in radians.

---

ap-appload-directory

Returns the directory for an application installed in the Appload Startup Suite.

`(ap-appload-directory AppName)`

Arguments

AppName　　　　　　A string, specifying the application file name and extension.

Returns

A string, specifying the directory in which the application is installed, or `nil`.

---

ap-asin

Returns the arcsine, in radians, of a number.

`(ap-asin number)`

Arguments

number　　　　　　　A number.

Returns

The arcsine, in radians, of *number*.

---

ap-block-insertable

Returns T if a block with the specified block name may be inserted into the drawing.

`(ap-block-insertable bname)`

Arguments

bname　　　　　　　A string, specifying the name of the block to query.

If the specified block is not currently defined, ap-block-insertable searches AutoCAD search path to locate a *.dwg* file with the same name as the block. If the file is found, ap-block-insertable uses the file to define the block.

Returns

T if the specified block name may be inserted into the drawing, `nil` otherwise.

ap-bitclear

Returns logical bitwise ~int1 AND int2.

```
(ap-bitclear int1 int2)
```

Arguments

int1   Integer specifying the bits to be cleared in *int2*.

int2   Integer

Returns

Returns logical bitwise ~*int1* AND *int2*.

---

ap-bitsetp

Returns T if the logical bitwise AND of a two of integers is non-zero.

```
(ap-bitsetp int1 int2)
```

Arguments

int1   Integer

int2   Integer

Returns

T if any bit in *int1* is also set in *int2*, nil otherwise.

---

ap-bulge->angle

Converts a polyline bulge number to an included angle in radians.

```
(ap-bulge->angle bulge)
```

Arguments

bulge   Number

Returns

A real number specifying the included angle in radians corresponding the to the bulge.

---

ap-callback-int

Sets the integer variable controlled by an edit box.

```
(ap-callback-int key value)
```

Arguments

key   String, specifying the tile and the name of the variable to be set.

value   String, specifying the value of the tile.

ap-callback-int uses the custom tile attributes as shown in Table A-2 to control the error checking for the tile. The error message (if any) is displayed in the tile named "error".

Remarks

The **ap-callback-int** function uses the custom dcl tile attributes as shown in Table A–2.

Table A–2: Custom Tile Attributes — ap-callback-int

Attribute	Usage
tag	The description of the tile to appear in error messages.
null_ok	Set to true to allow null values.
minus_ok	Set to true to accept negative values.
zero_ok	Set to true to accept zero values.

Returns

If successful, integer value of the tile, or "". `nil` otherwise.

ap-callback-list

Sets the list variable controlled by a list box or popup list.

```
(ap-callback-list key value)
```

Arguments

key	String, specifying the tile and name of the variable to be set.
value	String, specifying the value of the tile.

Returns

T, and sets the variable specified by *key* to a list of values extracted from the variable *key*_values.

Remarks

The **ap-callback-list** function uses the custom dcl tile attributes as shown in Table A–3

Table A–3: Custom Tile Attributes — ap-callback-list

Attribute	Usage
tag	The description of the tile to appear in error messages.
is_list	Set to true to identify a widget as a list box or popup list.
sort	Set to true to sort alphabetically the list values.

In addition, if the standard multiple_select attribute is *not* set to true, and the controlled variable is not one of the allowed values, it is set to the first value in the (possibly sorted) list.

ap-callback-radio

Sets the string variable controlled by a radio row or radio column.

```
(ap-callback-radio key value)
```

Arguments

key	String, specifying the radio row/column and name of the variable to be set.
value	String, specifying the name of the radio button pressed.

Returns

The value assigned to the named variable

ap-callback-real

Sets the real variable controlled by an edit box.

```
(ap-callback-real key value)
```

Arguments

key	String, specifying the tile and the name of the variable to be set.
value	String, specifying the value of the tile.

The **ap-callback-real** function uses the custom tile attributes as shown in Table A–4 to control the error checking for the tile. The error message (if any) is displayed in the tile named "error".

Table A–4: Custom Tile Attributes — ap-callback-real

Attribute	Usage
tag	The description of the tile to appear in error messages.
null_ok	Set to true to allow null values.
minus_ok	Set to true to accept negative values.
zero_ok	Set to true to accept zero values.
is_angle	Set to true to control angles expressed in radians.
units	Controls units display of numbers. Defaults to LUNITS or AUNITS.
prec	Controls precision display of numbers. Defaults to LUPREC or AUPREC.

Returns

If successful, real value of the tile, or "". `nil` otherwise.

---

ap-callback-string

Sets the string variable controlled by an edit box.

```
(ap-callback-string key value)
```

Arguments

key   String, specifying the tile and the name of the variable to be set.

value   String, specifying the value of the tile.

`ap-callback-string` uses the custom tile attributes as shown in Table A–5 to control the error checking for the tile. The error message (if any) is displayed in the tile named "error".

Table A–5: Custom Tile Attributes — ap-callback-string

Attribute	Usage
Tag	The description of the tile to appear in error messages.
Trim	Set to true to trim leading and trailing spaces.
null_ok	Set to true to accept blank entries.
spaces_ok	Set to true to accept internal spaces.

Returns

If successful, value of the tile. `nil` otherwise.

---

ap-callback-toggle

Sets the string variable controlled by a toggle.

```
(ap-callback-toggle key value)
```

Arguments

key   String, specifying the tile and the name of the variable to be set.

value   String, specifying the value of the tile.

Returns

"1" if the toggle is set, "0" if it is not.

ap-command-finish

Finishes the active AutoCAD command.

```
(ap-command-finish)
```

So long as the CMDACTIVE system variable is non-zero, **ap-command-finish** presses *ENTER* for you.

Returns

T

ap-degrees->radians

Converts an angle in degrees to an angle in radians.

```
(ap-degrees->radians angle)
```

Arguments

angle An angle in degrees

Returns

A real number, specifying the angle in radians.

ap-daysused

Returns the number a days an application has been in use.

```
(ap-daysused appname)
```

Arguments

appname A string, specifying the name of the application.

Returns

An integer.

Remarks

Returns 1 the first day called.

ap-deldict

Removes the specified symbol from the dictionary specified by the global variable AP:DICTNAME.

```
(ap-deldict symbol)
```

Arguments

symbol A symbol.

Global Variables

AP:DICTNAME A string, specifying the name of the dictionary.

Returns

T if the symbol was successfully deleted from the dictionary, nil otherwise.

ap-disassoc

Removes element(s) from an association list.

```
(ap-disassoc key alist all)
```

Arguments

key	Key of an element in an association list.
alist	An association list to be searched.
all	If nil, only the first entry for *key* will be removed.
	If non-nil, all entries of *key* will be removed.

Returns

The association list, with the one or all entries for *key* removed.

---

ap-done-dialog

Terminates a dialog box, and saves its current location.

(ap-done-dialog *status*)

Arguments

status   A positive integer that **start_dialog** will return.

You must call **ap-done-dialog** from within an action expression or callback function.

Global Variables

AP:DCL_LOCS
AP:DLG_STACK

Returns

A two-dimensional point list that is the (X,Y) location of the dialog box when the user exited it.

---

ap-done-dialog-conditional

Conditionally terminates a dialog box, and saves its current location.

(ap-done-dialog-conditional *vars prefix status*)

Arguments

vars   A list of the variables controlled by the dialog box. The same-named tile controls each variable.

prefix   A string, specifying the prefix to be assigned to each of the callback functions.

status   A positive integer that **start_dialog** will return

The ap-done-dialog-conditional function executes, in turn, the callback function for each of the variables in ***vars***.

Should any of the callback functions return nil, **ap-done-dialog-conditional** sets the focus of the dialog box to the offending tile, and does **not** terminate the dialog box.

If none of the callback functions returns nil, **ap-done-dialog-conditional** terminates the dialog box.

You must call **ap-done-dialog-conditional** from within an action expression or callback function.

Global Variables

AP:DCL_LOCS
AP:DLG_STACK

Returns

A two-dimensional point list that is the (X,Y) location of the dialog box when the user exited it.

---

ap-entlast

Returns the name of the last non-deleted entity (main entity or sub-entity) in the drawing.

(ap-entlast)

Returns

An entity name, or nil, if there are no entities in the current drawing.

Remarks

AutoLISP's **entlast** function returns the name of the last non-deleted main entity in a drawing.

3D polylines, 3D polygon meshes, polyface meshes, and blocks with attributes all have sub-entities.

The **ap-entlast** function returns the last main *or* sub-entity in the drawing.

---

ap-error

An error-handling function.

(ap-error *message*)

Arguments

message   A string containing a description of the error.

The **ap-error** function suppresses the display of "Function cancelled" and "quit / exit abort" error messages.

It then evokes **ap-pop-vars**, to restore system variables, **ap-pop-undo** to end undo-grouped operations, and **ap-pop-error** to restore the previous error handler.

Returns

No value is returned.

---

ap-extmax

Returns the upper-right corner of the extents of a list of points.

```
(ap-extmin points)
```

Arguments

points   A list of Points.

Returns

Returns the upper-right corner of the extents of *points*.

---

ap-extmin

Returns the lower-left corner of the extents of a list of points.

```
(ap-extmin points)
```

Arguments

points   A list of Points.

Returns

Returns the lower-left corner of the extents of *points*.

---

ap-exit-transparent

Forces the current application to quit, if there is a command active.

```
(ap-exit-transparent)
```

Prevents application from running when called transparently.

Returns

No value is returned.

---

ap-get-attributes

Returns a list of the Attributes attached to a Block Insertion.

```
(ap-get-attributes ename)
```

Arguments

ename   The entity name of the block insertion being queried.

Returns

A list of the Attributes, as returned by **entget**, attached to a Block Insertion.

---

ap-get-attribute-prompts

Returns a list of the tags and prompts for the Attributes attached to a Block Insertion.

```
(ap-get-attribute-prompts ename-bname)
```

Arguments

ename-bname   The entity name of the block insertion being queried.
                           OR
                     A string, specifying the name of the Block to be queried.

Returns

A list of the Attribute tags and prompts, as dotted pairs of strings, or `nil` if the Block is not defined.

---

ap-get-attribute-values

Returns a list of the tags and values of the Attributes attached to a Block Insertion.

`(ap-get-attribute-values ename)`

Arguments

`ename`   The entity name of the block insertion being queried.

Returns

A list of the Attribute tags and values, as dotted pairs of strings.

---

ap-get-authcode

Returns the authorization code stored for the application.

`(ap-get-authcode appname)`

Arguments

`appname`   A string, specifying the name of the application.

Returns

A string, specifying the authorization code for the application.

Remarks

If no authorization code is present, it's set to `"*DEMO*"`.

---

ap-get-ijth

Returns the jth element in the ith row of a matrix.

`(ap-get-ijth i j matrix)`

Arguments

`i, j`   Positive integers specifying the row and column of the matrix element to be retrieved. The first element in a row or column is at position 0.

`matrix`   A matrix

Returns

The ijth element of `matrix`.

Remarks

The first row is 0. The first column is 0.

---

ap-getcfg

Retrieves application data from the AppData/AP:APPNAME/symbol section of the *acad.cfg* file.

`(ap-getcfg symbol)`

Arguments

`symbol`   A symbol specifying the data to retrieve.

Global Variables

`AP:APPNAME`   A string, specifying the section of AppData from which to retrieve the symbol data.

Returns

Sets `symbol` to the value retrieved from the *acad.cfg* file, and returns that value.

Values may be of type `nil`, `integer`, `real`, `string`, or `symbol`.

ap-getdict

Retrieves LISP data from the drawing dictionary specified by the global variable AP:DICTNAME.

`(ap-getdict symbol)`

Arguments

`symbol`	A symbol.

Global Variables

`AP:DICTNAME`	A string, specifying the name of the dictionary.

Returns

Sets *symbol* to the value retrieved from the drawing dictionary, and returns that value.

ap-getdist

Pauses for user input of a distance with a default value.

`(ap-getdist pt msg default)`

Arguments

`pt`	A Point to be used as the base point in the current UCS. If *pt* isn't `nil`, the user is prompted for the second point.
`msg`	A string to be displayed to prompt the user.
`default`	A real value to be returned if the user presses ENTER, or `nil` for no default.

Returns

The distance specified by the user as a real number, or the value of *default*, if the user presses ENTER without entering a distance. If a 3D point is provided, the returned value is a 3D distance. However, setting the 64 bit of the `initget` function instructs **ap-getdist** to ignore the Z component of 3D points and to return a 2D distance.

ap-getdist honors the same `initget` functionality as does the **getdist** function.

ap-getint

Pauses for user input of an integer with a default value.

`(ap-getint msg default)`

Arguments

`msg`	A string to be displayed to prompt the user.
`default`	An integer value to be returned if the user presses ENTER, or `nil` for no default.

Returns

The integer value specified by the user, or the value of *default*, if the user presses ENTER without entering an integer.

ap-getint honors the same `initget` functionality as does the **getint** function.

ap-getkword

Pauses for user input of an keyword with a default value.

`(ap-getkword msg default)`

Arguments

`msg`	A string to be displayed to prompt the user.
`default`	A string value to be returned if the user presses ENTER, or `nil` for no default.

Returns

A string representing the keyword entered by the user, or the value of *default*, if the user presses ENTER without typing a keyword.

ap-getkword honors the same `initget` functionality as does the **getkword** function.

ap-getpoint

Pauses for user input of a point, and sets LASTPOINT to that point.

```
(ap-getpoint pt msg)
```

Arguments

`pt`	A 2D or 3D base point in the current UCS, or `nil`.
`msg`	A string to be displayed to prompt the user.

Returns

A 3D point, expressed in terms of the current UCS. The system variable LASTPOINT is set to the point entered by the user.

ap-getpoint honors the same **initget** functionality as does the **getpoint** function.

ap-getreal

Pauses for user input of a real number with a default value.

```
(ap-getreal msg default precision)
```

Arguments

`msg`	A string to be displayed to prompt the user.
`default`	A real value to be returned if the user presses ENTER, or `nil` for no default.
`precision`	An integer to override the system variable LUPREC, or `nil`

Returns

The real number specified by the user, or the value of `default`, if the user presses ENTER without entering a real number.

ap-getreal honors the same **initget** functionality as does the **getreal** function.

ap-getstring

Pauses for user input of a real number with a default value.

```
(ap-getstring cr msg default)
```

Arguments

`cr`	If not `nil`, this argument indicates that users can include blanks in their input string (and must terminate the string by pressing ENTER). Otherwise, the input string is terminated by space or by ENTER.
`msg`	A string to be displayed to prompt the user.
`default`	A string value to be returned if the user presses ENTER, or `nil` for no default.

Returns

The string specified by the user, or the value of `default`, if the user presses ENTER without entering a string. To enter an empty string without specifying the default, just enter ".".

ap-getstring honors the same **initget** functionality as does the **getstring** function.

ap-getyesno

Pauses for the user to input (or default) 'Yes' or 'No.'

```
(ap-getyesno msg default)
```

Arguments

`msg`	A string to be displayed to prompt the user. This typically will end with a question mark.
`default`	A string value of `"Yes"` or `"No"` to specify a default, or `nil` for no default.

Returns

T if the user responds or defaults to `"Yes"`, `nil` otherwise.

ap-init-vars

Initializes variables per the specified list of defaults.

```
(ap-init-vars defaults)
```

Arguments

defaults   An association list of variables and their default values.
                          Only those variables whose values are `nil` are modified.

Returns

T

Side Effects

Each of the enumerated variables whose values are `nil` is set to their default values.

ap-isonscreen

Verifies a user-specified point is visible in the current viewport.

```
(ap-isonscreen point)
```

Arguments

point   A Point in the current UCS.

Returns

T, if the point is on-screen, and `nil` if it is not.

ap-isplanview

Verifies the current view is plan in the current UCS.

```
(ap-isplanview)
```

Returns

T, if the current view is plan in the current UCS, and `nil` if it is not.

ap-item

Searches an association list for an element and returns the cdr of that association list entry.

```
(ap-item element alist)
```

Arguments

element   Key of an element in an association list.
alist   An association list to be searched.

Returns

The **cdr** of the alist entry, if successful. If **ap-item** does not find element as a key in alist, it returns nil.

ap-layer-ready

Thaws, unlocks, and turns on every layer in the drawing.

```
(ap-layer-ready)
```

Returns

A list containing the layer table prior to evoking ap-layer-ready. Each entry is as returned from **entget**.

Use **ap-layer-restore** to restore the previous layer settings.

ap-layer-restore

Restores the layers in a drawing per the provided layer table list.

```
(ap-layer-restore layers)
```

Arguments

A list containing the layer states to restore, as returned by **ap-layer-ready**. Each entry is as returned from **entget**.

Returns

A list as returned by (mapcar 'entmod *layers*)

---

ap-load-dialog

Loads a DCL file only if it is not currently loaded.

(ap-load-dialog *dclfile*)

Arguments

dclfile   A string that specifies the DCL file to load. If the *dclfile* argument does not specify a file extension, *.dcl* is assumed.

If not successful, an alert box is displayed.

Global Variables

AP:DCL_IDS   An association list of loaded *dclfile*s and their dcl_ids.

Returns

A positive integer value (dcl_id) if successful, or nil if **ap-load-dialog** can't open the file. The dcl_id is used as a handle in subsequent **ap-new-dialog**, **new_dialog**, and **unload_dialog** calls.

---

ap-log-base

Returns the logarithm of a number, in the specified base, as a real number.

(ap-log-base *base number*)

Arguments

base   A positive number representing the base of the desired logarithm

number   A positive number.

Returns

A real number

---

ap-matrix-adjoint

Returns the adjoint of the specified matrix.

(ap-matrix-adjoint *matrix*)

Arguments

matrix   A square matrix of numbers.

Returns

The adjoint of the specified matrix.

---

ap-matrix-cofactor

Returns the specified cofactor of the specified matrix.

(ap-matrix-cofactor *i j matrix*)

Arguments

i, j   Integers specifying the desired cofactor

matrix   A square matrix of numbers.

Returns

The determinant of the specified matrix with row i and column j removed, multiplied by $(-1)^{(i+j)}$.

ap-matrix-determinant

Returns the determinant of the specified matrix.

`(ap-matrix-determinant `*`matrix`*`)`

Arguments

matrix   A square matrix of numbers.

Returns

An integer if all the elements of *matrix* are integers.
A real if any elements of matrix are reals.

ap-matrix-dif

Returns the difference of two specified matrices.

`(ap-matrix-dif `*`matrix1`*` `*`matrix2`*`)`

Arguments

matrix1   A matrix of numbers.
matrix2   A matrix of numbers.

Returns

The matrix difference *matrix1 - matrix2*

ap-matrix-identity

Returns an identity matrix of the specified order.

`(ap-matrix-identity `*`order`*`)`

Arguments

order   A positive integer specifying the order (dimensions) of the identity matrix.

Returns

An identity matrix of the desired order.

ap-matrix-inverse

Returns the inverse of the specified matrix.

`(ap-matrix-inverse `*`matrix`*`)`

Arguments

matrix   A square matrix of numbers.

Returns

The inverse of the specified matrix, or `nil` if the matrix is singular.

ap-matrix-list-multiply

Returns the matrix product of the specified matrices.

`(ap-matrix-list-multiply `*`matrix_list`*`)`

Arguments

matrix_list   A list of matrices of numbers.

Returns

The matrix product of the specified matrices.

`    ((1.0 0.0) (0.0 1.0))`

ap-matrix-map

Returns a matrix of the result of executing a function with the individual elements of a matrix.

```
(ap-matrix-map function matrix)
```

Arguments

function	A function.
matrix	A matrix.

Returns

A matrix, with function applied to each element of the matrix.

Global Variables

i, j	Integers specifying the row and column of each element as *function* is evoked.

ap-matrix-multiply

Returns the matrix product of two specified matrices.

```
(ap-matrix-inverse matrix1 matrix2)
```

Arguments

matrix1	A matrix of numbers.
matrix2	A matrix of numbers.

Returns

The matrix product of the specified matrices.

ap-matrix remove

Removes the specified row and column from a matrix.

```
(ap-matrix-remove i j matrix)
```

Arguments

i, j	Positive integers specifying the row and column to be removed from the matrix.
matrix	A matrix.

Returns

The matrix, with row i and column j removed.

ap-matrix-rotation

Returns a transformation matrix with the specified rotations.

```
(ap-matrix-rotation xrot yrot zrot)
```

Arguments

xrot	A number specifying the angle of rotation, in radians, about the X-axis.
yrot	A number specifying the angle of rotation, in radians, about the Y-axis.
zrot	A number specifying the angle of rotation, in radians, about the Z-axis.

Returns

A transformation matrix with the specified rotations.

ap-matrix-scale

Returns the product of a scalar and a matrix.

```
(ap-matrix-scale scalar matrix)
```

Arguments

scalar	A number.
matrix	A matrix of numbers.

Returns

The product of *scalar* and *matrix*.

ap-matrix-scaling

Returns a transformation matrix with the specified scale factors.

(ap-matrix-scaling *xscale yscale zscale*)

Arguments

xscale	A number specifying the X scale factor of the transformation matrix.
yscale	A number specifying the Y scale factor of the transformation matrix.
zscale	A number specifying the z scale factor of the transformation matrix.

Returns

A transformation matrix with the specified scale factors.

ap-matrix-sum

Returns the sum of two specified matrices.

(ap-matrix-sum *matrix1 matrix2*)

Arguments

matrix1	A matrix of numbers.
matrix2	A matrix of numbers.

Returns

The matrix sum *matrix1 + matrix2*.

ap-matrix-translation

Returns a transformation matrix defined by two points.

(ap-matrix-translation *p1 p2*)

Arguments

p1, p2	3D points defining a translation vector.

Returns

A transformation matrix with the translation defined by the vector p1->p2.

ap-matrix-transpose

Returns the transpose of the specified matrix.

(ap-matrix-transpose *matrix*)

Arguments

matrix	A matrix of numbers.

Returns

The transpose of *matrix*.

ap-matrix-vector-product

Returns the product of the specified matrix and the specified vector.

(ap-matrix-vector-product *matrix vector*)

Arguments

matrix	A matrix of numbers.
vector	A vector of numbers.

Returns

The product of the *matrix* and *vector*.

ap-matrix-zero

Returns a zero matrix of the specified order.

```
(ap-matrix-zero order)
```

Arguments

order   A positive integer specifying the order (dimensions) of the zero matrix.

Returns

A zero matrix of the desired order.

ap-msgbox

Displays a VBA Message Box.

```
(ap-msgbox title message icon buttons defaultbutton)
```

Arguments

title   A string to be displayed in the title bar of the message box.

message   A string to be displayed as the message in the message box.

icon   One of the following:

apIconNone	Display no icon.
apIconCritical	Display Critical Message icon.
apIconQuestion	Display Warning Query icon.
apIconExclamation	Display Warning Message icon.
apIconInformation	Display Information Message icon.

buttons   One of the following:

apOKOnly	Display only the OK button only.
apOKCancel	Display OK and Cancel buttons.
apAbortRetryIgnore	Display Abort, Retry, and Ignore buttons.
apYesNoCancel	Display Yes, No, and Cancel buttons.
apYesNo	Display Yes and No buttons.
apRetryCancel	Display Retry and Cancel buttons.

defaultbutton   One of the following:

apDefaultButton1	First button is default.
apDefaultButton2	Second button is default.
apDefaultButton3	Third button is default.

Returns

An integer specifying the button pressed by the user. One of the following:

apOK	1	OK
apCancel	2	Cancel
apAbort	3	Abort
apRetry	4	Retry
apIgnore	5	Ignore
apYes	6	Yes
apNo	7	No

ap-midpoint

Returns the midpoint between two points.

```
(ap-midpoint point1 point2)
```

Arguments

point1, point2   Points.

Returns

The midpoint between two points.

ap-mod

Returns *number1* modulus *number2*.

```
(ap-mod number1 number2)
```

Arguments

number1	Number
number2	A positive number

Returns

A number. If either argument is a real, **ap-mod** returns a real, otherwise **ap-mod** returns an integer.

ap-mod divides *number1* by *number2*, and returns a remainder in the range $0 \leq remainder < number2$.

Remarks

The **ap-mod** function differs from the **rem** function insofar as the result is always a positive number.

ap-new-dialog

Begins a new dialog box and displays it at its previous location.

```
(ap-new-dialog dlgname dcl_id)
```

Arguments

dlgname	A string that specifies the dialog box.
dcl_id	The DCL file identifier obtained by **load_dialog** or **ap-load-dialog**.

Global Variables

AP:DCL_LOCS
AP:DLG_STACK

Returns

T, if successful, otherwise nil.

ap-polyline-segment

Returns detailed information about the polyline segment between two vertices.

```
(ap-polyline-segment vb1 vb2)
```

Arguments

vb1, vb2	Dotted pairs of vertex and bulge.

Returns

An association list with segment information.

ap-polyline-vblist

Returns, in the current UCS, a vertex-bulge list for the specified Polyline or LWPolyline.

```
(ap-polyline-vblist ename)
```

Arguments

ename	An entity name

Returns

A vertex-bulge list for the specified Polyline or LWPolyline, in the form ((vertex . bulge) ...)

Duplicate vertices are removed, and, if the Polyline or LWPolyline is closed, the last vertex duplicates the first.

The bulge is the tangent of 1/4 of the included angle for the arc between each selected vertex and the next vertex in the polyline's vertex list. A negative bulge value indicates that the arc goes clockwise from the selected vertex to the next vertex. A bulge of 0 indicates a straight segment, and a bulge of 1 is a semicircle.

ap-pop-command

Restores the values of BLIPMODE, HIGHLIGHT, OSMODE, and LASTPOINT from a frame on the AP:SYSVARS stack.

```
(ap-pop-command)
```

Global Variables

AP:SYSVARS   The system variable stack.

Returns

T

ap-pop-error

Restores the previous AutoLISP error handler from AP:OLD-ERROR.

```
(ap-pop-error)
```

Arguments

errorfunc   The desired error handling function.

Global Variables

AP:OLD-ERROR   Set to nil.

Returns

The old error handler.

ap-pop-frame

Restores system variables to their saved values, and clears the AP:SYSVARS stack up to the first frame marker.

```
(ap-pop-frame)
```

Global Variables

AP:SYSVARS   The system variable stack. All variables on the AP:SYSVARS stack are restored and removed up the first frame marker.

Returns

AP:SYSVARS after the frame has been popped.

ap-pop-undo

Restores the UNDOCTL system variable from AP:UNDOCTL.

```
(ap-pop-undo)
```

Global Variables

AP:UNDOCTL   Set to nil.

Returns

The new value of the UNDOCTL system variable.

ap-pop-vars

Restores system variables to their saved values, and clears the AP:SYSVARS stack.

```
(ap-pop-vars)
```

Global Variables

AP:SYSVARS   The system variable stack. Set to nil.

Returns

nil

ap-push-command

Saves the current values of BLIPMODE, HIGHLIGHT, OSMODE, and LASTPOINT as a frame on the AP:SYSVARS stack. Sets BLIPMODE, HIGHLIGHT, and OSMODE to 0.

```
(ap-push-command)
```

Global Variables

AP:SYSVARS   The system variable stack.

Returns

AP:SYSVARS after the command frame has been pushed.

ap-push-error

Saves the current AutoLISP error handler to AP:OLD-ERROR, and sets the error handler.

```
(ap-push-error errorfunc)
```

Arguments

errorfunc   The desired error handling function.

Global Variables

AP:OLD-ERROR   Set to the prior error handler.

Returns

errorfunc

ap-push-frame

Saves system variables to a frame on the AP:SYSVARS stack, and sets the system variables.

```
(ap-push-frame varlist)
```

Arguments

varlist   An association list of system variables with their new values. **ap-pop-frame** will restore the system variables to their saved values.

Global Variables

AP:SYSVARS   The system variable stack.

Returns

AP:SYSVARS after the command frame has been pushed.

ap-push-undo

Saves the current state of the UNDOCTL system variable to AP:UNDOCTL.

```
(ap-push-vars undo_required)
```

Arguments

undo_required   If non-nil, assures that Undo All is turned on.

Global Variables

AP:UNDOCTL   Saves the prior value of the UNDOCTL system variable.

Returns

The new value of the UNDOCTL system variable.

ap-push-vars

Saves system variables to the AP:SYSVARS stack, and sets the system variables.

```
(ap-push-vars varlist)
```

Arguments

varlist An association list of system variables with their new values. **ap-pop-vars** will restore the system variables to their saved values.

Global Variables

AP:SYSVARS The system variable stack.

Returns

T

---

ap-putcfg

Writes application data to the AppData/*AP:APPNAME*/*symbol* section of the *acad.cfg* file.

(ap-putcfg *symbol*)

Arguments

symbol A quoted symbol specifying the data to write, of type nil, integer, real, string, or symbol.

Global Variables

AP:APPNAME A string, specifying the section of AppData to which to write the *symbol* value.

Returns

The value of the symbol as a string.

---

ap-putdict

Stores LISP data in the drawing dictionary specified by the global variable AP:DICTNAME.

(ap-putdict *symbol*)

Arguments

symbol A symbol.

Global Variables

AP:DICTNAME A string, specifying the name of the dictionary.

Returns

The value of *symbol*.

---

ap-radians->degrees

Converts an angle in radians to an angle in degrees.

(ap-radians->degrees *angle*)

Arguments

angle A number

Returns

A real number.

---

ap-rationalize

Converts a number to a rational fraction.

(ap-rationalize *number*)

Arguments

number A number.

Returns

Returns a list containing the numerator and denominator of the fraction.

ap-rdc

Returns a list containing all but the last element of the specified list.

`(ap-rdc list)`

Arguments

`list`   A list.

Returns

Returns a list containing all but the last element of the specified list.

---

ap-replace

Replaces an element in an association list.

`(ap-replace element alist)`

Arguments

`element`   An association list element to appear in the association list `alist`.

`alist`   An association list.

Returns

The association list, with the element replacing the corresponding element in the `alist`. If there is no corresponding element, it is appended to the end of the list.

---

ap-remove-nth

Removes the nth element in a list.

`(ap-remove-nth n list)`

Arguments

`n`   Positive integer specifying the list element to be replaced.
                     The first element in the list is at position 0.

`list`   A list.

Returns

The list, with the nth element removed.

---

ap-reset-vars

Sets each of the specified variables to the specified values.

`(ap-reset-vars saved_values)`

Arguments

`saved_values`   An association list of variables and their values, as returned from **ap-set-vars**.

Returns

T

---

ap-restore-var

Restores the specified system variable from the AP:SYSVARS stack.

`(ap-restore varname)`

Arguments

`varname`   A string, specifying the name of the system variable to restore.

Global Variables

`AP:SYSVARS`   The system variable stack.

Returns

The restored value of the system variable, or `nil`, if it wasn't saved on the AP:SYSVARS stack.

ap-rotate

Moves the head of the list to the end of a list.

```
(ap-rotate list)
```

Arguments

list   A list.

Returns

The list, with the head of the list moved to the end of the list.

ap-round

Rounds a number the nearest integral multiple of another.

```
(ap-round number base)
```

Arguments

number   A number to be rounded.

base   A number specifying the base of the rounding.

Returns

Returns the integral multiple of *base* closest to *number*.

ap-save-vars

Returns a list of variables with their values.

```
(ap-save-vars defaults)
```

Arguments

defaults   A list of variables.

Returns

An association list of variables with their values. This list may be used with **ap-reset-vars** to restore the variables to their saved values.

ap-set-attribute-values

Updates the Attributes attached to a Block insertion.

```
(ap-set-attribute-values ename values)
```

Arguments

ename   The entity name of the block insertion being updated.

values   A list of dotted pairs of tags and values.

Returns

T if any attributes were modified, nil otherwise.

ap-set-authcode

Returns sets authorization code stored for the application.

```
(ap-get-authcode appname code)
```

Arguments

appname   A string, specifying the name of the application.

code   A string, specifying the authorization code.

Returns

A string.

Remarks

If no authorization code is present, it is set to "*DEMO*".

ap-set-ijth

Replaces the jth element in the ith row of a matrix.

```
(ap-set-ijth i j matrix newelement)
```

Arguments

i, j	Positive integers specifying the row and column of the matrix element to be replaced. The first element in a row or column is at position 0.
matrix	A matrix.
newelement	New value for the ijth element of *matrix*.

Returns

The matrix, with *newelement* in the desired position.

ap-set-nth

Replaces the nth element in a list.

```
(ap-set-nth n list newelement)
```

Arguments

n	Positive integer specifying the list element to be replaced. The first element in the list is at position 0.
list	A list.
newelement	New value for the nth element of *list*.

Returns

The list, with *newelement* in the desired position.

Remarks

If n is not in the range $0 \leq n <$ `(length alist))`, the list is unchanged.

ap-set-tiles

Sets the tiles in a dialog box to the values of the variables with the same names as the tiles.

```
(ap-set-tiles params)
```

Arguments

params   A list of the AutoLISP variables to be displayed in the tiles.

ap-set-tiles sets edit boxes for reals, integers, strings, distances, and angles; list boxes, popdown lists, radio rows, radio columns, text boxes, and toggles.

Returns

T

ap-sgn

Returns the sign of the first argument times the second.

```
(ap-sgn number1 number2)
```

Arguments

number1, number2 Numbers

Returns

The second number multiplied by the sign of the first number.

ap-sort

Sorts the elements in a list according to a given compare function.

```
(ap-sort list compare-function)
```

Arguments

list   Any list.

compare-function   A comparison function. This can be any function that accepts two arguments and returns T (or any non-nil value) if the first argument precedes the second in the sort order. The *compare-function* value can take one of the following forms:

A symbol (function name)

'(lambda (a1 a2) ...)

(function (lambda (a1 a2) ...))

Returns

A list containing the elements of list in the order specified by comparison-function. Duplicate elements are not eliminated from the list.

---

ap-sqr

Returns the square of the specified number.

(ap-sqr *number*)

Arguments

number   Number any number.

Returns

The square of the specified number.

---

ap-sscopy

Copies a selection set.

(ap-copy *ss*)

Arguments

ss   A selection set or nil.

Returns

A selection set containing the objects in *ss*, or nil.

---

ap-sslength

Returns an integer containing the number of objects (entities) in a selection set.

(ap-sslength *ss*)

Arguments

ss   A selection set or nil.

Returns

An integer. Does not fail if *ss* is nil.

Remarks

The standard AutoLISP **sslength** function fails if called with a null selection set. The **ap-sslength** function returns 0.

---

ap-sslist

Returns a list of the entity names in a selection set.

(ap-sslist *ss*)

Arguments

ss   A selection set or nil.

Returns

A list of the entity names in a selection set.

---

ap-ssprevious

Establishes the specified selection set as the previous selection set.

```
(ap-ssprevious ss)
```

Arguments

ss   A selection set or nil.

---

ap-string-equal

Compares two strings for equality without regard to case.

```
(ap-string-equal string1 string2)
```

Arguments

string1, string2 Strings

Returns

T, if the arguments are equal without regard to case, nil otherwise.

---

ap-string-less

Compares two strings for without regard to case.

```
(ap-string-less string1 string2)
```

Arguments

string1, string2 Strings

Returns

T, if *string1* is less than *string2*, without regard to case.

---

ap-string-search

Searches for the specified pattern in a string, ignoring case.

```
(ap-string-search pattern string start-pos)
```

Arguments

pattern	A string containing the pattern to be searched for.
string	The string to be searched for *pattern*.
start-pos	An integer identifying the starting position of the search. The first character of the string is position 0.

Returns

An integer representing the position in the string where the specified pattern was found, or nil if the pattern is not found; the first character of the string is position 0.

Unlike the **vl-string-search** function, **ap-string-search** is case-insensitive.

---

ap-string-subst

Substitutes all occurrences of one string for another, within a string.

```
(ap-string-search new-string old-string string)
```

Arguments

new-string	The string to replace *old-string*.
old-string	The string to be replaced by new-string.
string	The string to be searched for *old-string*.

Returns

The value of *string* after any substitutions have been made.

Unlike the **vl-string-subst** function, **ap-string-subst** replaces all occurrences of *old-string* in *string*.

---

ap-stringp

Verifies that an item is a string.

```
(ap-stringp item)
```

Arguments

item   An AutoLISP expression.

Returns

T if *item* evaluates to a string, nil otherwise.

---

ap-symbol-name

Returns a string containing the lowercase name of a symbol.

```
(ap-symbol-name symbol)
```

Arguments

symbol   Any AutoLISP symbol.

Returns

A string containing the name of the supplied *symbol* argument, in lowercase.

---

ap-swap

Swaps the values of the specified symbols.

```
(ap-swap sym1 sym2)
```

Arguments

sym1, sym2   The symbols to be swapped.

Returns

T

---

ap-tan

Returns the tangent of an angle.

```
(ap-tan angle)
```

Arguments

angle   An angle, in radians.

Returns

A real number representing the tangent of *angle*.

---

ap-tile-exists

Verifies that a tile exists in the active dialog box.

```
(ap-tile-exists tile)
```

Arguments

tile   A key string that specifies the tile. This parameter is case-sensitive.

Returns

T if *tile* exists in the active dialog box, nil otherwise.

ap-timer

Returns the elapsed time in seconds since the last call to ap-timer.

```
(ap-timer)
```

Global Variables

AP:LASTTIME   Stores the date and time ap-timer was last called, as a Julian date and fraction in a real number:

Returns

A real number indicating the elapsed time in seconds from the last call to **ap-timer,** or nil if never called.

ap-transform-point

Transforms a 3D point, or list of 3D points, by a transform matrix.

```
(ap-transform-point matrix point(s))
```

Arguments

matrix   Transformation matrix.

point(s)   A 3D point, or list of 3D points.

Returns

A 3D point, or list of 3D points, transformed by the transformation matrix.

ap-transform-vector

Transforms a 3D vector, or list of 3D vectors, by a transform matrix.

```
(ap-transform-vector matrix vector(s))
```

Arguments

matrix   Transformation matrix.

vector(s)   A 3D vector, or list of 3D vectors.

Returns

A 3D vector, or list of 3D vectors, transformed by the transformation matrix.

ap-truncate

Truncates a number the nearest smaller integral multiple of another.

```
(ap-truncate number base)
```

Arguments

number   A number to be truncated.

base   A number specifying the base of the truncation.

Returns

Returns the integral multiple of base closest to and less than number.

ap-update-var

Updates a system variable saved in the AP:SYSVARS stack.

```
(ap-update varname value)
```

Arguments

varname   A string, specifying the system variable to be updated in AP:SYSVARS.

value   A new value for the variable.

Global Variables

AP:SYSVARS

Returns

The new value, if successful, or `nil` if the specified variable is not in the AP:SYSVARS stack.

---

ap-vector-dif

Returns the difference of two specified vectors.

`(ap-vector-dif vector1 vector2)`

Arguments

`vector1`	A vector (list) of numbers.
`vector2`	A vector (list) of numbers.

Returns

The vector difference `vector1 - vector2`.

---

ap-vector-cross-product

Returns the dot product two specified vectors.

`(ap-vector-cross-product vector1 vector2)`

Arguments

`vector1`	A vector (list) of numbers.
`vector2`	A vector (list) of numbers.

Returns

The vector cross product `vector1 × vector2`.

---

ap-vector-dot-product

Returns the dot product two specified vectors.

`(ap-vector-dot-product vector1 vector2)`

Arguments

`vector1`	A vector (list) of numbers.
`vector2`	A vector (list) of numbers.

Returns

The vector dot product `vector1 • vector2`.

---

ap-vector-normal

Returns a unit vector normal to the plane defined by p1 p2 p3.

`(ap-vector-normal p1 p2 p3)`

Arguments

`p1, p2, p3`	Points on the plane.

Returns

A unit vector perpendicular to the plane, or `nil` if the points do not define a plane.

---

ap-vector-magnitude

Returns the magnitude of a vector.

`(ap-vector-magnitude vector)`

Arguments

`vector`	A vector (list) of numbers.

Returns

The magnitude of *vector* as a real number .

---

ap-vector-scale

Returns the product of a scalar and a vector.

(ap-vector-scale *scalar vector*)

Arguments

scalar	A number
vector	A vector (list) of numbers.

Returns

The product of *scalar* and *vector*.

---

ap-vector-sum

Returns the sum of two specified vectors.

(ap-vector-sum *vector1 vector2*)

Arguments

vector1	A vector (list) of numbers.
vector2	A vector (list) of numbers.

Returns

The vector sum *vector1* + *vector2*.

---

ap-vector-unit

Returns a unit vector in the direction of *vector*.

(ap-vector-unit *vector*)

Arguments

vector	A vector (list) of numbers.

Returns

A unit vector in the direction of vector, or nil the magnitude of *vector* is zero.

---

ap-vslide

Displays a slide in the specified image tile or image button.

(ap-vslide *tile slide*)

Arguments

tile	A string, specifying the name of the tile in which to display the slide.
slide	The name of the slide file, or slide within a slide library, to display.

Returns

nil

The appearance of a particular slide depends on three attributes defined for each image tile: border, border_color, and background.

The border attribute specifies the width of a frame (in pixels) to be drawn about the slide. The default is to display no border (border = 0).

The border_color and background attributes specify the colors for the border and the slide background, respectively. They default to the current background of the AutoCAD graphics screen.

Colors may be specified with AutoCAD color numbers (ACI), or by one of the logical color numbers shown in Table A–6.

Table A–6: Image Tile Colors

Color number	Meaning
0	Current background of the AutoCAD graphics screen. Slides will appear in the dialog box just as they do with the VSLIDE command.
–15	Current dialog box background color
–16	Current dialog box foreground color (for text)
–18	Current dialog box line color

ap-waswere

A string with the appropriate suffix for *integer*.

(ap-waswere *integer singular plural*)

Arguments

integer	An integer.
singular	A suffix to be added if *integer* = 1.
plural	A suffix to be added if *integer* ≠ 1.

Returns

A string with the appropriate suffix for *integer*.

ap-zero

Returns 0 if the absolute value of the argument is $<10^{-15}$.

(ap-zero *number*)

Arguments

number	A number.

Returns

Returns 0 if the absolute value of *number* is $<10^{-15}$, otherwise, returns *number*.

Artful Programming API ActiveX Utility Functions

apax-2d-point

Creates an ActiveX-compatible 2D point structure.

(apax-2d-point *Point*)

Arguments

Point	A 2D or 3D point.

Returns

A 2D point structure.

apax-3d-point

Creates an ActiveX-compatible 3D point structure.

(apax-3d-point *Point*)

Arguments

Point	A 2D or 3D point.

Returns

A 3D point structure.

apax-boolean->lisp

Converts an ActiveX Automation Boolean to an AutoLISP compatible value.

(apax-boolean->lisp *Item*)

Arguments

Item   An ActiveX Automation Boolean value.

Returns

T if *Item* = **:vlax-true**, nil otherwise.

Remarks

Use this function whenever an ActiveX Automation function returns **:vlax-true** or **:vlax-false.**

apax-dxf-join

Joins a safearray of integer codes and a safearray of variants into a list of DXF data.

(apax-dxf-join *dxfTypes dxfValues*)

Arguments

dxfTypes   A safearray of integer codes.
dxfValues   A safearray of variants.

Returns

A list of DXF data.

apax-dxf-split

Splits list of dxf data into a safearray of integer codes and a safearray of variants.

(apax-dxf-split *dxfList* '*dxfTypes* '*dxfValues*)

Arguments

dxfList   List of dxf data.

Returns

The list *(dxfTypes dxfData).*
dxfTypes   Safearray of integers from the *dxfList.*
dxfValues   Safearray of variants from the *dxfList.*

Remarks

dxfTypes and *dxfValues* must be quoted symbols.

apax-find-template

Returns the fully qualified path to a template file.

(apax-find-template *FileName*)

Arguments

FileName   String, specifying the template file to find.

Returns

The fully qualified path to a template file, or nil if not found.

apax-get-UserSpace

Returns the ModelSpace or PaperSpace object in which the user is drawing.

(apax-get-UserSpace *Object*)

Arguments

Object   The Document object to be queried.

Returns

ModelSpace or PaperSpace object in which the user is drawing.

---

apax-init

Initializes Artful Programming object variables

```
(apax-init)
```

Returns

T

Remarks

The apax-init function initializes a number of AutoLISP variables to some useful objects, as shown in Table A–7.

Table A–7: AutoLISP variables set by apax-init

Variable	Object
apAcadApp	(vlax-get-acad-object)
apActiveDimStyle	apThisDrawing.ActiveDimStyle
apActiveDocument	ApAcadApp.ActiveDocument
apActiveLayer	apThisDrawing.ActiveLayer
apActiveLayout	apThisDrawing.ActiveLayout
apActiveLinetype	apThisDrawing.ActiveLinetype
apActiveSelectionSet	apThisDrawing.ActiveSelectionSet
apActiveSpace	apThisDrawing.ActiveSpace
apActiveTextStyle	apThisDrawing.ActiveTextStyle
apActiveUCS	apThisDrawing.ActiveUCS
apActiveViewport	apThisDrawing.ActiveViewport
apAppPath	ApAcadApp.Path
apBlocks	apThisDrawing.Blocks
apCaption	ApAcadApp.Caption
apDatabase	apThisDrawing.Database
apDatabasePreferences	apThisDrawing.Preferences
apDictionaries	apThisDrawing.Dictionaries
apDimStyles	apThisDrawing.DimStyles
apDocuments	ApAcadApp.Documents
apElevationModelSpace	apThisDrawing.ElevationModelSpace
apElevationPaperSpace	apThisDrawing.ElevationPaperSpace
apFullName	ApAcadApp.FullName
apFullName	apThisDrawing.FullName
apGroups	apThisDrawing.Groups

apHeight	ApAcadApp.Height
apLayers	apThisDrawing.Layers
apLayouts	apThisDrawing.Layouts
apLinetypes	apThisDrawing.Linetypes
apLocaleId	ApAcadApp.LocaleId
apMenuBar	ApAcadApp.MenuBar
apMenuGroups	ApAcadApp.MenuGroups
apModelSpace	apThisDrawing.ModelSpace
apName	ApAcadApp.Name
apName	apThisDrawing.Name
apPaperSpace	apThisDrawing.PaperSpace
apPath	apThisDrawing.Path
apPlot	apThisDrawing.Plot
apPlotConfigurations	apThisDrawing.PlotConfigurations
apPreferences	ApAcadApp.Preferences
apPreferencesDisplay	ApAcadApp.Preferences.Display
apPreferencesDrafting	ApAcadApp.Preferences.Drafting
apPreferencesFiles	ApAcadApp.Preferences.Files
apPreferencesOpenSave	ApAcadApp.Preferences.OpenSave
apPreferencesOutput	ApAcadApp.Preferences.Output
apPreferencesProfiles	ApAcadApp.Preferences.Profiles
apPreferencesSelection	ApAcadApp.Preferences.Selection
apPreferencesSystem	ApAcadApp.Preferences.System
apPreferencesUser	ApAcadApp.Preferences.User
apRegisteredApplications	apThisDrawing.RegisteredApplications
apSelectionSets	apThisDrawing.SelectionSets
apTextStyles	apThisDrawing.TextStyles
apThisDrawing	apApAcadApp.ActiveDocument
apUserCoordinateSystems	apThisDrawing.UserCoordinateSystems
apUtility	apThisDrawing.Utility
apVBE	ApAcadApp.VBE
apVersion	ApAcadApp.Version
apViewports	apThisDrawing.Viewports
apViews	apThisDrawing.Views

apax-IsModelSpace

Determines if user is drawing in ModelSpace.

(apax-IsModelSpace *Object*)

Arguments

Object   The Document object to be queried.

Returns

T if user is drawing in ModelSpace, nil otherwise.

---

apax-IsSimpleBlock

Determines if specified Block object is neither an Xref nor a Layout.

(apax-IsSimpleBlock *Object*)

Object   The Block object to be queried.

Returns

T if the Block object is neither an Xref nor Xref Dependent nor a Layout, nil otherwise.

---

apax-lisp->boolean

Converts an AutoLISP value to an ActiveX Automation Boolean.

(apax-lisp->boolean *Value*)

Arguments

Value   An AutoLISP value.

Returns

Returns **:vlax-false** if item is nil, **:vlax-true** otherwise.

---

apax-lisp->variant

Converts an AutoLISP value to an ActiveX Automation variant safearray.

(apax-lisp->variant *Value*)

Arguments

Value   An integer, real, string, object, or variant, or list of the same, to be converted into a variant.

Returns

An ActiveX Automation variant.

---

apax-list->safearray

Converts an AutoLISP list to an ActiveX Automation safearray.

(apax-list->safearay *List*)

Arguments

List   A list of integers, reals, strings, objects, or variants to be converted into a safearray.

Returns

An ActiveX Automation safearray.

---

apax-list->variant

Converts an AutoLISP list to an ActiveX Automation variant safearray.

(apax-list->variant *List*)

Arguments

List   A list of integers, reals, strings, objects, or variants to be converted into a variant safearray.

Returns

An ActiveX Automation variant safearray.

---

apax-points->variant

Converts a list of points to a ActiveX Automation compatible structure.

(apax-points->variant *List*)

Arguments

List   A list of 2D or 3D points.

Returns

A variant safearray of doubles.

Remarks

It is up to the programmer to assure the points are 2D or 3D as required by the receiving ActiveX Automation function.

---

apax-variant->2d-points

Converts a variant safearray of doubles to a list of 2D points.

(apax-variant->2d-points *vArray*)

Arguments

vArray   The variant safearray of doubles to be converted.

Returns

A list of 2D points.

---

apax-variant->3d-points

Converts a variant safearray of doubles to a list of 3D points.

(apax-variant->3d-points *vArray*)

Arguments

vArray   The variant safearray of doubles to be converted.

Returns

A list of 3D points.

---

apax-variant->list

Converts a variant safearray to a list.

(apax-variant->list vArray)

Arguments

vArray   The variant safearray to be converted.

Returns

A list of elements of the safearray.

---

apax-variant->lisp

Converts a variant into an AutoLISP-compatible value.

(apax-variant->lisp *Variant*)

Arguments

Variant   Variant to be converted.
                         Variant safearrays are converted to lists.
                         Variant Booleans are converted to T or nil
                         Other variants are returned as their values.

Returns

An AutoLISP-compatible value.

```
_$ (setq var (apax-lisp->variant '("A" "B" "C")))
#<variant 8200 ...>
_$ (apax-variant->lisp var)
("A" "B" "C")
_$ (setq var (apax-lisp->variant 2))
#<variant 2 2>
_$ (apax-variant->lisp var)
2
```

Artful Programming API Reactor Functions

apr-attach-reactors

Attaches editor and mouse reactors to user-defined callback functions.

```
(apr-attach-reactors AppName)
```

Arguments

AppName   String, specifying the prefix for user-defined callback functions for the events shown in Table A–8.

Returns

T

Remarks

For each event of which you wish to be notified, you create a callback function with the name **appname-event**, where **appname** is the name of your application, and **event** is the name of the event.

Table A–8: Artful Programming Callback Events

Event Name	Description	Command List
abortDxfIn	The DXF import was aborted.	*nil.*
abortDxfOut	The DXF export was aborted.	*nil.*
beginClose	The drawing is about to be closed.	*nil.*
beginDoubleClick	The user double-clicked.	*Point clicked*
beginDwgOpen	A drawing is about to be opened.	*(filename)*
beginDxfIn	A DXF import was invoked.	*nil.*
beginDxfOut	A DXF export was invoked	*nil.*
beginRightClick	The user right-clicked.	*Point clicked*
beginSave	The drawing is about to be saved.	*(default-filename)*
commandCancelled	The AutoCAD command was canceled.	*(command-name)*
commandEnded	The AutoCAD command ended.	*(command-name)*
commandFailed	An AutoCAD command failed.	*(command-name)*
commandWillStart	An AutoCAD command was invoked.	*(command-name)*
databaseToBeDestroyed	The drawing is about to be closed without saving changes.	*nil.*

dwgFileOpened	A new drawing has been opened.	`(filename)`
dxfInComplete	The DXF import completed.	`nil.`
dxfOutComplete	The DXF export completed.	`nil.`
endDwgOpen	A drawing has been opened..	`(filename)`
lispCancelled	The evaluation of an AutoLISP expression was canceled.	`nil.`
lispEnded	The evaluation of an AutoLISP expression ended.	`nil.`
lispWillStart	An AutoLISP expression is about to be evaluated.	`(AutoLISP expression)`
saveComplete	The drawing has been saved.	`(filename)`
sysVarChanged	A system variable has been changed.	`(sysvar-name flag)` `1 if successful,` `0 if not.`
sysVarWillChange	A system variable is about to be change	`(sysvar-name)`
unknownCommand	An unknown command was issued.	`(command-name)`

apr-remove-all-reactors

Removes all mouse, database, editor, linker, and object reactors.

```
(apr-remove-all-reactors)
```

Returns

T

Remarks

A most useful function while debugging applications.

apr-remove-reactors

Removes application-specific mouse, database, editor, linker, and object reactors.

```
(apr-remove-reactors AppName)
```

Arguments

AppName   String, specifying the prefix for user-defined callback functions for the events shown in Table A-8.

Returns

T

Remarks

This function belongs in every beginClose callback function.

apr-trace

Prints callback function parameters.

```
(apr-Trace Routine TraceList)
```

Arguments

Routine   A quoted Symbol or String to identify the calling routine.

TraceList   A list to be printed, along with *Routine*.

Returns

T

Remarks

This function belongs in every beginClose callback function.

Artful Programming API ActiveX Method Functions

Activate Method

Makes the specified drawing active.

```
(apa-Activate Object)
```

Arguments

Object   The document object to activate.

Returns

T

Remarks

Under AutoCAD 2000, an AutoLISP program in runs only in the active drawing. For this reason, the **ap-activate** function does **not** return a value **until** the current drawing is re-activated.

Add Method

Creates an object and adds it to the appropriate collection.

```
(apa-Add Object Name)
```

Arguments

Object   The Dictionaries, Dimstyles, Documents, Groups; Layers, Layouts, Linetypes, PopupMenus, RegisteredApps, SelectionSets, TextStyles, Toolbars, Views, Viewports, or collection to which to add the new object.

Name   A string, specifying the name of the new object.
                       For the Documents collection, this name is optional, and represents the name of the drawing template to use and will accept a URL address or a filename.

Returns

The newly added object.

Add Method — Blocks

Creates a block (definition) object and adds it to the appropriate collection.

```
(apa-AddBlock Object Name InsertionPoint)
```

Arguments

Object   The Blocks collection to which to add the new object.
Name   A string, specifying the name of the new object.
InsertionPoint   A point in the WCS specifying the insertion point of the block.

Returns

The newly added object.

Add Method — Hyperlinks

Creates a hyperlink object and adds it to the appropriate collection.

```
(apa-AddHyperlink Object URL Description Location)
```

Arguments

Object	The collection to which to add the new object.
URL	A string, specifying the URL of the new object.
Description	A string, specifying the description of the hyperlink to add, or `nil`.
Location	A given location, such as a named view in AutoCAD or a bookmark in a word processing program. If you specify a named view to jump to in an AutoCAD drawing, AutoCAD restores that view when the hyperlink is opened.

Returns

The newly added hyperlink object.

Add Method — PlotConfigurations

Creates a PlotConfiguration object and adds it to the appropriate collection.

`(apa-AddPlotConfiguration Object Name ModelSpace)`

Arguments

Object		The collection to which to add the new object.
Name		A string, specifying the name of the new object.
ModelSpace	non-nil	The plot configuration applies only to the Model Space tab.
	nil	The plot configuration applies to all layouts.

Returns

The newly added object.

Add Method — UCSs

Creates a UCS object and adds it to the appropriate collection.

`(apa-AddUCS Object Name Origin XAxisPoint YAxisPoint)`

Arguments

Object	The Blocks collection to which to add the new object.
Name	A string, specifying the name of the new object.
Origin	A point specifying the origin of the new UCS.
XAxisPoint	A point on the positive portion of the X Axis of the UCS.
YAxisPoint	A point on the positive portion of the Y Axis of the UCS.

Returns

The newly added object.

Add3Face Method

Creates a 3DFace object given four vertices.

`(apa-Add3DFace Object Point1 Point2 Point3 Point4)`

Arguments

Object	ModelSpace Collection, PaperSpace Collection, or Block object in which to create graphical object.
Point1-4	Four points, each specifying a vertex of the 3DFace .

Returns

The newly created 3DFace object.

Add3DMesh Method

Creates a free-form M×N 3D mesh, given M, N, and list of points in row order.

`(apa-Add3DMesh Object M N Points)`

Arguments

Object	ModelSpace Collection, PaperSpace Collection, or Block object in which to create graphical object.
M, N	The integer dimensions of the point array.
Points	A list of *M×N* 3D points. The vertices for row *M* must be supplied before the vertices for row *M+1*.

Returns

The newly created PolygonMesh object.

Remarks

Use **apa-set-MClose** to close the mesh in the *M* direction, or **apa-set-NClose** to close the mesh in the *N* direction.

Add3DPoly Method

Creates a 3D polyline from the given list of points.

(apa-Add3DPoly *Object Points*)

Arguments

Object	ModelSpace Collection, PaperSpace Collection, or Block object in which to create graphical object.
Points	A list of 3D points.

Returns

The newly created 3DPoly object.

Remarks

Use **apa-set-Closed** to close the polyline.

AddArc Method

Creates an arc given the center, radius, start angle, and end angle of the arc.

(apa-AddArc *Object Center Radius StartAngle EndAngle*)

Arguments

Object	ModelSpace Collection, PaperSpace Collection, or Block object in which to create graphical object.
Center	Point specifying the center of the arc.
Radius	Number specifying the Radius of the arc.
StartAngle	Number specifying the start angle of the arc, in radians.
EndAngle	Number specifying the end angle of the arc, in radians.

Returns

The newly created Arc object.

Remarks

Arcs are always drawn counter-clockwise.

AddAttribute Method

Creates an attribute definition at the given location with the specified properties.

(apa-AddAttribute *Object Height Mode Prompt InsertionPoint Tag Value*)

Arguments

Object	ModelSpace Collection, PaperSpace Collection, or Block object in which to create graphical object.

Height	Number specifying text height.
Mode	The sum of any combination of the following constants:
	acAttributeModeNormal
	acAttributeModeInvisible
	acAttributeModeConstant
	acAttributeModeVerify
	acAttributeModePreset
Prompt	String, specifying the user prompt.
InsertionPoint	Point specifying the insertion point for the attribute definition.
Tag	String, specifying the name of the attribute.
Value	String, specifying the default attribute value.

Returns

The newly created Attribute object.

AddBox Method

Creates a 3D solid box with edges parallel to the axes of the WCS.

(apa-AddBox *Object Origin Length Width Height*)

Arguments

Object	ModelSpace Collection, PaperSpace Collection, or Block object in which to create graphical object.
Origin	Point specifying the center of the bounding box. This is *not* a corner of the box.
Length	Positive number specifying the length (X) of the box.
Width	Positive number specifying the width (Y) of the box.
Height	Positive number specifying the height (Z) of the box.

Returns

A 3DSolid object as the newly created box.

AddCircle Method

Creates a circle given a center point and radius parallel to the XY plane of the WCS.

(apa-AddCircle *Object Center Radius*)

Arguments

Object	ModelSpace Collection, PaperSpace Collection, or Block object in which to create graphical object.
Center	Point specifying the center of the circle.
Radius	Positive number specifying the radius of the circle.

Returns

The newly created Circle object.

AddCone Method

Creates a 3D solid cone with the base parallel to the XY plane of the WCS.

(apa-AddCone *Object Center Radius Height*)

Arguments

| Object | ModelSpace Collection, PaperSpace Collection, or Block object in which to create graphical object. |
| Center | Point specifying the center of the bounding box. This is *not* center of the base. |

Radius	Positive number specifying the radius of the base of the cone.
Height	Positive number specifying the height of the cone.

Returns

A 3DSolid object as the newly created cone.

AddCylinder Method

Creates a 3D solid cylinder with the base parallel to the XY plane of the WCS.

(apa-AddCylinder *Object Center Radius Height*)

Arguments

Object	ModelSpace Collection, PaperSpace Collection, or Block object in which to create graphical object.
Center	Point specifying the center of the bounding box. This is *not* center of the base.
Radius	Positive number specifying the radius of the base of the cylinder.
Height	Positive number specifying the height of the cylinder.

Returns

A 3DSolid object as the newly created cylinder.

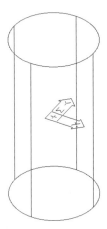

AddDim3PointAngular Method

Creates an 3-point angular dimension.

```
(apa-AddDim3PointAngular Object Vertex FirstEndPoint SecondEndPoint
      TextPoint)
```

Arguments

Object	ModelSpace Collection, PaperSpace Collection, or Block object in which to create graphical object.
Vertex	Point specifying the angle vertex.
FirstEndPoint	Point specifying the first angle endpoint.
SecondEndPoint	Point specifying the second angle endpoint.
TextPoint	Point at which the dimension text should be placed.

Returns

The newly created Dim3PointAngular object.

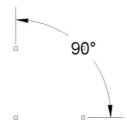

AddDimAligned Method

```
(apa-AddDimAligned Object FirstEndPoint SecondEndPoint TextPoint)
```
Creates an aligned dimension.

Arguments

Object	ModelSpace Collection, PaperSpace Collection, or Block object in which to create graphical object.
FirstEndPoint	Point specifying the first extension line origin.
SecondEndPoint	Point specifying the second extension line origin.
TextPoint	Point at which the dimension text should be placed.

Returns

The newly created DimAligned object.

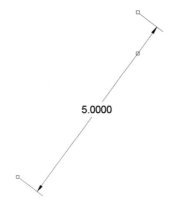

AddDimAngular Method

Creates an two-line angular dimension.

```
(apa-AddDimAngular Object FirstStartPoint FirstEndPoint SecondStartPoint
        SecondEndPoint TextPoint)
```

Arguments

Object	ModelSpace Collection, PaperSpace Collection, or Block object in which to create graphical object.
FirstStartPoint	Point specifying the first line startpoint.
FirstEndPoint	Point specifying the first line endpoint.
SecondStartPoint	Point specifying the second line startpoint.
SecondEndPoint	Point specifying the second line endpoint.
TextPoint	Point at which the dimension text should be placed.

Returns

The newly created DimAngular object.

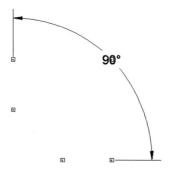

AddDimDiametric Method

Creates a diametric dimension for a circle or arc given the two points on the diameter.

```
(apa-AddDimDiametric Object Chordpoint FarChordpoint LeaderLength)
```

Arguments

Object	ModelSpace Collection, PaperSpace Collection, or Block object in which to create graphical object.
Chordpoint	Point specifying the first diameter point on the circle or arc.
FarChordpoint	Point specifying the second diameter point on the circle or arc.
LeaderLength	Positive number specifying the distance between Chordpoint and the dimension text or dogleg, or 0 for a default leader.

Returns

The newly created DimDiametric object.

Ø1.2500

AddDimOrdinate Method

Creates an ordinate dimension given the definition point, and the leader endpoint.

(apa-AddDimOrdinate *Object DefinitionPoint LeaderEndpoint UseXAxis*)

Arguments

Object	ModelSpace Collection, PaperSpace Collection, or Block object in which to create graphical object.
DefinitionPoint	Point specifying point to be dimensioned.
LeaderEndpoint	Point specifying the endpoint of the leader.
UseXAxis	non-nil Creates an X-Datum dimension.
	nil Creates a Y-Datum dimension.

Returns

The newly created DimOrdinate object.

AddDimRadial Method

Creates a radial dimension for a circle or arc given the center point and a point on the circle or arc.

(apa-AddDimRadial *Object Center ChordPoint LeaderLength*)

Arguments

Object	ModelSpace Collection, PaperSpace Collection, or Block object in which to create graphical object.
Center	Point specifying the center of the circle or arc.
ChordPoint	Point specifying the a point on the circle or arc.
LeaderLength	Positive number specifying the distance between ChordPoint and the dimension text or dogleg, or 0 for a default leader.

Returns

The newly created DimRadial object.

AddDimRotated Method

Creates a rotated dimension.

(apa-AddDimRotated *Object FirstEndPoint SecondEndPoint TextPoint
        RotationAngle*)

Arguments

Object	ModelSpace Collection, PaperSpace Collection, or Block object in which to create graphical object.
Vertex	Point specifying the angle vertex.
FirstEndPoint	Point specifying the first extension line origin.
SecondEndPoint	Point specifying the second extension line origin.
TextPoint	Point at which the dimension text should be placed.
RotationAngle	Number specifying the rotation angle of the dimension in radians.

Returns

The newly created DimRotated object.

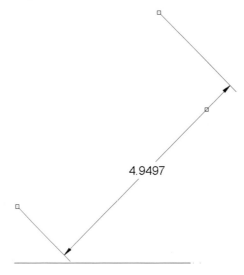

4.9497

AddEllipse Method

Creates an ellipse parallel to the XY plane of the WCS given the center point, an endpoint on the major axis, and the Axis ratio.

`(apa-AddEllipse Object Center MajorAxis AxisRatio)`

Arguments

Object	ModelSpace Collection, PaperSpace Collection, or Block object in which to create graphical object.
Center	Point specifying the center of the ellipse.
MajorAxis	Point specifying an endpoint of the major axis of the ellipse, *relative to the center of the ellipse.*
AxisRatio	A number expressing the minor / major axis ratio. `0 < AxisRatio ≤ 1`

Returns

A newly created Ellipse object.

Remarks

To draw an elliptical arc, draw an ellipse, then use the **apa-set-StartAngle** and **apa-set-EndAngle** functions to modify the ellipse.

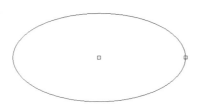

AddEllipticalCone Method

Creates a 3D solid elliptical cone with the base parallel to the XY plane of the WCS.

`(apa-AddEllipticalCone Object Center XRadius YRadius Height)`

Arguments

Object	ModelSpace Collection, PaperSpace Collection, or Block object in which to create graphical object.
Center	Point specifying the center of the bounding box. This is *not* center of the base.
XRadius	Positive number specifying the X radius of the base of the cone.
YRadius	Positive number specifying the Y radius of the base of the cone.
Height	Positive number specifying the height of the cone.

Returns

A 3DSolid object as the newly created elliptical cone.

Remarks

There is no restriction that *XRadius* be the major radius of the elliptical base.

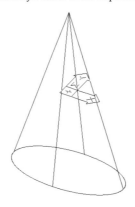

AddEllipticalCylinder Method

Creates a 3D solid elliptical cylinder with the base parallel to the XY plane of the WCS.

`(apa-AddEllipticalCylinder Object Center XRadius YRadius Height)`

Arguments

Object	ModelSpace Collection, PaperSpace Collection, or Block object in which to create graphical object.
Center	Point specifying the center of the bounding box. This is *not* center of the base.

XRadius	Positive number specifying the X radius of the base of the cylinder.
YRadius	Positive number specifying the Y radius of the base of the cylinder.
Height	Positive number specifying the height of the cylinder.

Returns

A 3DSolid object as the newly created elliptical cylinder.

Remarks

There is no restriction that *XRadius* be the major radius of the elliptical base.

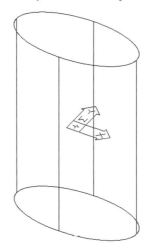

AddExtrudedSolid Method

Creates an extruded solid given the Profile, Height, and TaperAngle.

(apa-AddExtrudedSolid *Object Profile Height TaperAngle*)

Arguments

Object	ModelSpace Collection, PaperSpace Collection, or Block object in which to create graphical object.
Profile	Region object representing the profile to be extruded.
Height	Number specifying the height of the extrusion along the Z-axis of the profile's coordinate system. Positive numbers will extrude in the positive Z direction; negative numbers, the negative Z direction.
TaperAngle	Number specifying the taper angle in radians. Positive angles taper in from the base, negative angles taper out.

$$-p/2 < TaperAngle < p/2$$

Returns

A 3DSolid object as the newly created extruded solid, or nil if not successful.

Remarks

The original objects are *not* deleted.

You can extrude only 2D planar regions.

Taper angles and/or extrusion heights that would result in a self-intersecting solid will fail.

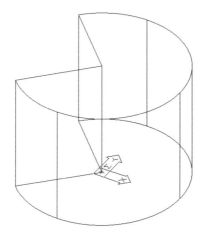

AddExtrudedSolidAlongPath Method

Creates an extruded solid given the profile and an extrusion path.

`(apa-AddExtrudedSolidAlongPath Object Profile Path)`

Arguments

Object	ModelSpace Collection, PaperSpace Collection, or Block object in which to create graphical object.
Profile	Region object representing the profile to be extruded.
Path	Polyline, Circle, Ellipse, Spline, or Arc object.

Returns

A 3DSolid object as the newly created extruded solid, or `nil` if not successful.

Remarks

The original objects are *not* deleted.

You can extrude only 2D planar regions.

The path cannot lie on the same plane as the profile.

Paths and/or profiles that would result in a self-intersecting solid will fail.

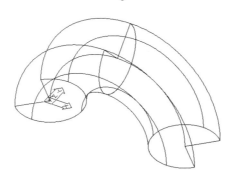

AddFitPoint Method

Adds the fit point to the spline at a given index.

`(apa-AddFitPoint `*`Object Index FitPoint`*`)`

Arguments

Object	The Spline object to which this method applies.
Index	Position in the fit point list to add the fit point. If the index is a negative number, then the point is added to the beginning of the spline. If the index exceeds the number of fit points in the spline, then the point is added to the end of the spline.
FitPoint	A point.

Returns

T if successful, `nil` otherwise.

Remarks

AutoCAD adds the point and refits the spline through the new set of points. To view the changes, use **apa-update** or **apa-Regen**.

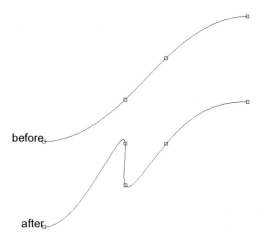

AddHatch Method

Creates a Hatch object.

`(apa-AddHatch `*`Object PatternType PatternName Associativity`*`)`

Arguments

Object	ModelSpace Collection, PaperSpace Collection, or Block object in which to create graphical object.
PatternType	One of the following values: **acHatchPatternTypePredefined** **acHatchPatternTypeUserDefined** **acHatchPatternTypeCustomDefined**
PatternName	String, specifying the name of the hatch pattern.
Associativity	non-`nil` creates associative hatch. `nil` creates non-associative hatch.

Returns

The newly created Hatch object.

Remarks

After the Hatch object has been created using the AddHatch method, you must define the outer hatch boundary with the AppendOuterLoop method.

After your outer loops have been defined, you can define islands with the AppendInnerLoop method.

To display the hatch, you must use the Evaluate method.

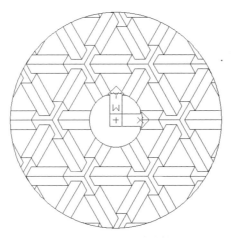

AddItems Method

Adds one or more objects to the specified selection set.

(apa-AddItems *Object Items*)

Arguments

Object	SelectionSet Object.
Items	A list of objects (not entity names) to be added to the selection set.

Returns

T

AddLeader Method

Creates a Leader object, given a list of points and an annotation object.

(apa-AddLeader *Object Points Annotation LeaderType*)

Arguments

Object	ModelSpace Collection, PaperSpace Collection, or Block object in which to create graphical object.
Points	List of points.
Annotation	Tolerance, MText, or BlockRef object.
LeaderType	One of the following values: **acLineNoArrow** **acLineWithArrow** **acSplineNoArrow** **acSplineWithArrow**

Returns

The newly created Leader object.

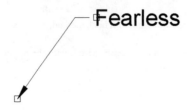

Fearless

AddLightweightPolyline Method

Creates a lightweight polyline from the given list of points in the OCS.

`(apa-AddLightweightPolyline` *`Object Points`*`)`

Arguments

Object	ModelSpace Collection, PaperSpace Collection, or Block object in which to create graphical object.
Points	List of OCS Coordinates. 3D points will have their Z-component ignored.

Returns

The newly created LightweightPolyline object.

Remarks

For more information about the OCS, search the AutoCAD on-line documentation for OCS.

If 3D points are provided, the Z-values are ignored. The elevation for the polyline will be set at the current elevation for the layout.

Use **apa-set-Closed** to close the polyline.

To create an arc segments, first create the polyline, then add bulge to the individual segments you want to be arcs via the **apa-SetBulge** function.

AddLine Method

Creates a line between two points.

`(apa-AddLine` *`StartPoint EndPoint`*`)`

Arguments

StartPoint	Point specifying the startpoint of the line.
EndPoint	Point specifying the endpoint of the line.

Returns

The newly created Line object.

AddMenuItem Method

Adds a popup menu item to a popup menu.

`(apa-AddMenuItem` *`Object Index Label Macro`*`)`

Arguments

Object	The PopupMenu object in which to add the newly created object.
Index	Either an integer or a string, specifying the position in which to add the popup menu object.
Label	A string, specifying the label to appear in the menu.
Macro	A string, specifying the menu macro to add.

Returns

The newly created PopupMenuItem Object.

AddMInsertBlock Method

Inserts an array of blocks parallel to the XY plane of the UCS.

```
(apa-AddMinsertBlock Object Name InsertionPoint XScale YScale ZScale
        RotationAngle NumRows NumCols RowSpacing ColumnSpacing)
```

Arguments

Object	The ModelSpace, PaperSpace, or Block collection object in which to add the newly created object.
Name	String, specifying the name of the AutoCAD drawing file or the name of the block to insert. If it is a file name, include any path information necessary for AutoCAD to find the file and the .dwg extension.
InsertionPoint	Point specifying the insertion point of the block, in the WCS.
XScale	Non-zero number specifying the X scale factor of the block.
YScale	Non-zero number specifying the Y scale factor of the block.
ZScale	Non-zero number specifying the Z scale factor of the block.
RotationAngle	Number specifying the rotation angle in radians for the block.
NumRows	A positive integer representing the number of rows for the array.
NumColumns	A positive integer representing the number of columns for the array.
RowSpacing	A number specifying the distance between the array rows.
ColumnSpacing	A number specifying the distance between the array columns.

Returns

The newly created BlockRef object, or `nil` if not successful.

Remarks

Searches the AutoCAD Search Path if the specified block is not in defined in the current drawing.

AddMLine Method

Creates a multiline defined by a list of points.

```
(apa-AddMLine Object Points)
```

Arguments

Object	The ModelSpace Collection, PaperSpace Collection, or Block object in which to create graphical object.
Points	List of Points.

Returns

The newly created MLine object.

AddMText Method

Creates an MText entity in a rectangle defined by the insertion point and the width of the bounding box.

```
(apa-AddMText Object InsertionPoint Width Text)
```

Arguments

Object	ModelSpace Collection, PaperSpace Collection, or Block object in which to create graphical object.
InsertionPoint	Point specifying the insertion point of the bounding box.
Width	Number specifying the width of the bounding box.
Text	The text string for the MText object.

Returns

The newly created MText object.

AddObject Method

Adds an object to a named dictionary.

```
(apa-AddObject Object Keyword ObjectName)
```

Arguments

Object	Dictionary
	The dictionary object to which this method applies.
Keyword	String, specifying the keyword to be listed in the dictionary for this object.
ObjectName	String; input-only.
	The rxClassName of the object to be created in the dictionary.

Returns

The newly created object.

AddPoint Method

```
(apa-AddPoint Object Point)
```
Creates a Point object at a given location.

Arguments

Object	ModelSpace Collection, PaperSpace Collection, or Block object in which to create graphical object.
Point	Point specifying the location of the new point.

Returns

The newly created Point object.

AddPolyfaceMesh Method

Creates a polyface mesh from a list of vertices and a list of faces.

```
(apa-AddPolyfaceMesh Object Vertices Faces)
```

Arguments

Object	ModelSpace Collection, PaperSpace Collection, or Block object in which to create graphical object.
Vertices	List of 3D points specifying the vertices of the polyface mesh.
Faces	List of faces specifying the faces of the polyface mesh. Each face is a list of exactly four integers, each integer specifying one of the vertices in the list of vertices.

Returns

The newly created PolyfaceMesh object.

Remarks

The minimum number of vertices is 4, the maximum is 32767.

The minimum number of faces is 1, the maximum is 32767.

The first vertex is number 1. A negative integer specifies an invisible edge. A vertex number of 0 may be used as the fourth corner of a triangular face.

AddPolyline Method

Creates a polyline from the given list of points in the current OCS.

```
(apa-AddPolyline Object Points)
```

Arguments

Object	ModelSpace Collection, PaperSpace Collection, or Block object in which to create graphical object.
Points	List of OCS Coordinates. 3D points will have their Z-components ignored.

Returns

The newly created Polyline object.

Remarks

For more information about the OCS, search the AutoCAD on-line documentation for OCS.

If 2D points are provided, the Z-values are set to zero.

Use `apa-set-Closed` to close the polyline.

To create an arc segments, first create the polyline, then add bulge to the individual segments you want to be arcs via the `apa-SetBulge` function.

This method exists for backward compatibility only. Use the AddLightweightPolyline method to create polylines with an optimized format that saves memory and disk space.

AddPViewport Method

Adds a paper space viewport, given the center, height, and width.

(apa-AddPViewport *Object Center Width Height*)

Arguments

Object	PaperSpace Collection object in which to create the viewport.
Center	Point specifying the center of the viewport.
Width	Positive number specifying the width of the viewport.
Height	Positive number specifying the height of the viewport.

Returns

The newly created PViewport Object.

AddRaster Method

Creates a new raster image based on an existing image file.

(apa-AddRaster *Object FileName InsertionPoint ScaleFactor RotationAngle*)

Arguments

Object	The ModelSpace Collection, PaperSpace Collection, or Block object to which this method applies.
FileName	A string, specifying the path and file name of the image.
InsertionPoint	A WCS point specifying where the raster image will be created.
ScaleFactor	A non-zero positive number specifying the raster image scale factor.
RotationAngle	A number specifying the rotation angle in radians for the raster image.

Returns

The newly created Raster object.

Remarks

Searches the AutoCAD Search path.

AddRay Method

Creates a ray object passing through the specified points.

(apa-AddRay *Object Point1 Point2*)

Arguments

Object	The ModelSpace Collection, PaperSpace Collection, or Block object to which this method applies.
Point1	A WCS point specifying the finite start point of the ray.
Point2	A WCS point specifying a point through which the ray will pass. The ray extends from Point1, through Point2 to infinity.

Returns

The newly created Ray object.

AddRegion Method

Creates regions from a list of objects.

`(apa-AddRegion Object ObjectList)`

Arguments

Object	ModelSpace Collection, PaperSpace Collection, or Block object in which to create graphical object.
ObjectList	A list of objects to be made into regions. The original objects **not** are deleted.

Returns

A list of the newly created Region objects.

Remarks

This function creates a region out of each closed planar loop defined by the ObjectList. If loops share common endpoints, the resulting regions may be unpredictable, if not remarkable.

The original objects are **not** deleted.

AddRevolvedSolid Method

Creates a revolved solid.

`(apa-AddRevolvedSolid Object Profile AxisPoint AxisDir RevAngle)`

Arguments

Object	ModelSpace, PaperSpace, or Block collection object in which to add the newly created object.
Profile	Region object representing the profile to be extruded.
AxisPoint	Point specifying a point on the axis of revolution, in the WCS.
AxisDir	Vector specifying the direction of the axis of revolution, in the WCS.
RevAngle	Number representing the angle of revolution in radians. Use 2p for a full circle of revolution.

Returns

A 3DSolid object as the newly created revolved solid.

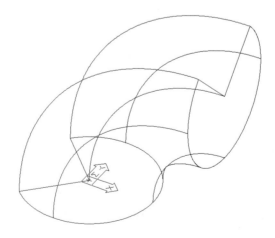

AddSeparator Method

Adds a separator to an existing menu or toolbar.

(apa-AddSeparator *Object Index*)

Arguments

Object	The PopupMenu or Toolbar object in which to add the newly created object.
Index	Either an integer or a string, specifying the position in which to add the popup menu object.

Returns

The newly created separator object.

AddShape Method

Creates a Shape object, parallel to the XY plane of the UCS.

(apa-AddShape *Object Name InsertionPoint ScaleFactor RotationAngle*)

Arguments

Object	ModelSpace, PaperSpace, or Block collection object in which to add the newly created object.
Name	String, specifying the name of the shape.
InsertionPoint	Point specifying where the shape will be placed, in the WCS.
ScaleFactor	Positive number specifying the shape scale factor.
RotationAngle	Number specifying the rotation angle in radians for the shape.

Returns

The newly created Shape object.

Remarks

The file containing the desired shape must be loaded using the using the LoadShapeFile method before adding a shape.

AddSolid Method

Creates a 2D solid-filled polygon.

(apa-AddSolid *Object Point1 Point2 Point3 Point4*)

Arguments

Object	ModelSpace, PaperSpace, or Block collection object in which to add the newly created object.
Point1-Point4	Points specifying the vertices of the polygon, in the WCS.

Returns

The newly created Solid object.

AddSphere Method

Creates a sphere given the center and radius.

(apa-AddSphere *Object Center Radius*)

Arguments

Object	ModelSpace, PaperSpace, or Block collection object in which to add the newly created object.
Center	Point specifying the center of the sphere, in the WCS.
Radius	Positive Number specifying the radius of the sphere.

Returns

A 3DSolid object as the newly created sphere.

AddSpline Method

Creates a quadratic or cubic NURBS curve.

(apa-AddSpline *Object Points StartTangent EndTangent*)

Arguments

Object	ModelSpace, PaperSpace, or Block collection object in which to add the newly created object.
Points	List of Points defining the spline curve, in the WCS.
StartTangent	Vector specifying the tangency of the curve at the first point, in the WCS.
EndTangent	Vector specifying the tangency of the curve at the last point, in the WCS.

Returns

The newly created Spline object.

Remarks

Splines are created open. Use **apa-set-Closed** to close the spline.

AddSubMenu Method

Adds a submenu to an existing menu.

(apa-AddSubMenu *Object Index Label*)

Arguments

Object	The PopupMenu object in which to add the newly created object.
Index	Either an integer or a string, specifying the position in which to add the popup menu object.
Label	A string, specifying he label for the menu item. The label may contain DIESEL string expressions. Labels also specify the accelerator keys (keyboard key sequences) that correspond to the menu item by placing an ampersand (&) in front of the accelerator character.

Returns

The newly created PopupMenu object.

AddText Method

Creates a single line of text, parallel to the XY plane of the UCS.

(apa-AddText *Object TextString InsertionPoint Height*)

Arguments

Object	ModelSpace, PaperSpace, or Block collection object in which to add the newly created object.
TextString	String, specifying the text to be displayed.
InsertionPoint	Point specifying where the text will be placed, in the WCS.
Height	Positive Number specifying the height of the text.

Returns

The newly created Text object.

AddTolerance Method

Creates a tolerance entity, parallel to the XY plane of the UCS.

(apa-AddTolerance *Object TextString InsertionPoint Direction*)

Arguments

Object	ModelSpace, PaperSpace, or Block collection object in which to add the newly created object.
TextString	String, specifying the text to be displayed.
InsertionPoint	Point specifying where the tolerance symbol will be placed, in the WCS.
Direction	Vector specifying the direction of the tolerance symbol in the WCS.

Returns

The newly created Tolerance object.

AddToolbarButton Method

Adds a toolbar item to a toolbar at a specified position.

```
(apa-AddToolbarButton Object Index Name HelpString Macro IsFlyout)
```

Arguments

Object	The Toolbar object in which to add the newly created object.
Index	Either an integer or a string, specifying the position in which to add the popup menu object.
Name	A string, specifying the name of the toolbar button. This string is displayed as the ToolTip when the cursor is placed over the toolbar button.
HelpString	The string to appears in the AutoCAD status line for the button.
Macro	The button macro for this button.
IsFlyout	If non-`nil`, specifies the button is to be a flyout button.

Returns

The newly created toolbar item.

Remarks

The *Name* argument must be comprised of alphanumeric characters with no punctuation other than a dash (-) or an underscore (_).

You cannot have more than one button with the same toolbar.

You add or remove toolbar buttons only when the toolbar is visible.

AddTorus Method

Creates a torus at the given location, parallel to the XY plane of the WCS.

```
(apa-AddTorus Object Center TorusRadius TubeRadius)
```

Arguments

Object	ModelSpace, PaperSpace, or Block collection object in which to add the newly created object.
Center	Point specifying the center of the torus, in the WCS.
TorusRadius	Positive number specifying the distance from the center of the torus to the center of the tube.
TubeRadius	Positive number specifying the radius of the tube.

Returns

A 3DSolid object as the newly created torus.

AddTrace Method

Creates a 2D solid-filled polygon, parallel to the XY plane of the UCS.

```
(apa-AddTrace Object Point1 Point2 Point3 Point4)
```

Arguments

Object	ModelSpace, PaperSpace, or Block collection object in which to add the newly created object.
Point1-Point4	Points specifying the vertices of the polygon, in the WCS.

Returns

The newly created Trace object.

Remarks

You probably should be using **apa-AddLightweightPolyline** instead of this function.

AddVertex Method

Adds a vertex to a lightweight polyline.

(apa-AddVertex *Object Index Point*)

Arguments

Object	LightweightPolyline object to which to add vertex.
Index	Integer specifying the index of vertex to be added, starting at 0.
Point	Point specifying the coordinates of the vertex to be added, in the OCS.

Returns

T

Remarks

If the index specifies an existing vertex, the new vertex will be inserted *before* the existing vertex.

AddWedge Method

Creates a 3D solid box with edges parallel to the axes of the WCS.

(apa-AddWedge *Object Center Length Width Height*)

Arguments

Object	ModelSpace, PaperSpace, or Block collection object in which to add the newly created object.
Center	Point specifying the center of the bounding box, and as it so happens, the center of the oblique face of the wedge, in the WCS. This is *not* a corner of the wedge.
Length	Positive number specifying the length (X) of the wedge.
Width	Positive number specifying the width (Y) of the wedge.
Height	Positive number specifying the height (Z) of the wedge.

Returns

A 3DSolid object as the newly created wedge.

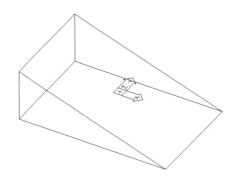

AddXLine Method

Creates an xline passing through two specified points.

(apa-AddXLine *Object Point1 Point2*)

Arguments

Object	ModelSpace, PaperSpace, or Block collection object in which to add the newly created object.
Point1	Point, in the WCS.
Point2	Point, in the WCS.

Returns

The newly created XLine object.

AppendInnerLoop Method

Appends an inner loop to the hatch.

(apa-AppendInnerLoop *Object Loop*)

Arguments

Object	The Hatch object to which to add the inner loop.
Loop	List of Line, LightweightPolyline, Polyline, Circle, Ellipse, Spline, and Region objects forming a properly closed boundary.

Returns

T

AppendItems Method

Appends one or more entities to the specified group.

(apa-AppendItems *Object Objects*)

Arguments

Object	The Group object to which to append the specified objects.
Objects	A list of objects to be appended to the Group Object.

AppendOuterLoop Method

Appends an inner loop to the hatch.

(apa-AppendInnerLoop *Object Loop*)

Arguments

Object	The Hatch object to which to add the inner loop.
Loop	List of Line, LightweightPolyline, Polyline, Circle, Ellipse, Spline, and Region objects forming a properly closed boundary.

Returns

T

AppendVertex Method

Appends a vertex to a 3DPoly or Polyline object;
Appends a row of vertices to a PolygonMesh object.

Arguments

(apa-AppendVertex *Object Point*)

Object	3DPoly or Polyline object to which to append vertex.
Points	Point specifying the coordinates of the vertex to be appended, in the WCS.

```
(apa-AppendVertex Object Points)
```
Object   PolygonMesh object to which to append a row of vertices.

Points   A list of Points specifying the coordinates of a row of vertices to be appended.

Returns

T

ArrayPolar Method

Creates a polar array of objects, parallel to the XY plane of the UCS.

```
(apa-ArrayPolar Object CenterPoint NumObjects AngleToFill)
```

Arguments

Object   The Drawing object (or list of objects) to which this method applies.

CenterPoint   Point specifying the center point for the polar array, in the WCS.

NumObjects   Positive integer, greater than 1, specifying the *total* number of objects in the polar array, including the original object. The number of objects created is *NumObjects-1*.

AngleToFill   Number specifying the angle to fill in radians; positive for counterclockwise rotation, negative for clockwise.

Returns

A list of the newly created objects.

Remarks

This method does not support the 'Rotate arrayed objects' option of AutoCAD's ARRAY command.

The reference point used depends on the type of object being arrayed; e.g., the center point of a circle or arc, the insertion point of a block or shape, the start point of text.

Adding or removing members from a collection, while iterating through it with the **vlax**-for function, may cause an error.

ArrayRectangular Method

Creates a 2D or 3D rectangular array of objects, parallel to the XY plane of the UCS.

```
(apa-ArrayRectangular Object NumRows NumColumns NumLevels DistBetRows
        DistBetColumns DistBetLevels)
```

Arguments

Object   The Drawing object (or list of objects) to which this method applies.

NumRows   Positive integer specifying the number of rows in the array.

NumColumns   Positive integer specifying the number of columns in the array.

NumLevels   Positive integer specifying the number of levels in the array.

DistBetRows   Number specifying the distance between rows.

DistBetColumns Number specifying the distance between columns.

DistBetLevels Number specifying the distance between levels.

Returns

A list of the newly created objects.

Remarks

Unlike AutoCAD's ARRAY command, this method ignores the current snap rotation angle.

Adding or removing members from a collection, while iterating through it with the **vlax**-for function, may cause an error.

You cannot use this method on AttributeReference objects.

AttachExternalReference Method

Attaches an xref the drawing, parallel to the XY plane of the UCS.

```
(apa-AttachExternalReference Object PathName XrefName InsertionPoint XScale
        YScale ZScale RotationAngle Overlay)
```

Object	ModelSpace, PaperSpace, or Block collection object in which to add the newly created object.
PathName	String, specifying the path name of the external reference. The AutoCAD search path is searched.
XrefName	String, specifying the name of the xref to be created.
InsertionPoint	Point specifying the insertion point of the xref, in the WCS.
XScale	Non-zero number specifying the *X* scale factor of the xref.
YScale	Non-zero number specifying the *Y* scale factor of the xref.
ZScale	Non-zero number specifying the *Z* scale factor of the xref.
RotationAngle	Number specifying the rotation angle in radians for the xref.
Overlay	non-nil The xref instance is an overlay.
	nil The xref instance is an attachment.

Returns

The newly created ExternalReference object.

AuditInfo Method

Evaluates the integrity of the drawing.

```
(apa-AuditInfo Object FixError)
```

Arguments

Object	The Drawing object to which this method applies
FixError	non-nil Attempt to fix errors
	nil Do not attempt to fix errors.

Returns

T

Remarks

If no errors are detected, no error report file is created.

Bind Method

Binds an xref to a drawing.

```
(apa-Bind Object PrefixNames)
```

Arguments

Object	The block object to which this method applies.
PrefixNames	non-nil Symbol names prefixed with *<blockname>x*.
	nil Symbol names are not prefixed.

Returns

T

Boolean Method

Performs a Boolean operation between two 3DSolid or Region objects.

```
(apa-Boolean FirstObject Operation SecondObject)
```

Arguments

FirstObject	3DSolid or Region object to be modified.
Operation	Integer specifying the operation to be performed.

                               acUnion: Performs an union operation.

                               acIntersection: Performs an intersection operation.

                               acSubtraction: Performs a subtraction operation.

SecondObject	3DSolid or Region object to be Unioned, Intersected, or Subtracted from the first solid.

Returns

T

Remarks

Whether or not the operation results in a null solid, the first object is modified, and the second object deleted.

CheckInterference Method

Checks for interference between two solids and, if specified, creates a solid from the interference.

`(apa-CheckInterference FirstObject SecondObject CreateSolid)`

Arguments

FirstObject	First 3DSolid object to be checked for interference.	
SecondObject	Second 3DSolid object to be checked for interference.	
CreateSolid	non-nil	Creates the interference solid.
	nil	Does not create the interference solid.

Returns

The newly created 3DSolid object, or T, or `nil`.

Clear Method

Clears the specified selection set of all items.

`(apa-Clear Object)`

Arguments

Object	The SelectionSet object to be cleared.

Remarks

Clearing a Selection Set has no effect on the (now) former members of the Selection Set. They're not removed from the drawing, just from the Selection Set.

ClipBoundary Method

Specifies the clipping boundary for a raster image.

`(apa-ClipBoundary Object Points)`

Arguments

Object	The Raster object to which the boundary should be applied.
Points	A list of Points specifying the clipping boundary.

Returns

T

Remarks

If 3D points are provided, the Z-values will be ignored.

Copy Method

Duplicates the given object.

`(apa-Copy Object)`

Arguments

`Object`   The Drawing object (or list of objects) to be duplicated.

Returns

The newly created duplicate object (or list of objects).

Remarks

Adding or removing members from a collection, while iterating through it with the **vlax**-for function, may cause an error.

You cannot use this method on AttributeReference objects.

CopyObjects Method

Duplicates multiple objects (deep cloning).

`(apa-CopyObjects` *Object Objects NewOwner*`)`

Arguments

`Object`   The Database (xref) or Document object to which this method applies.

`Objects`   List of primary objects to be copied. All the objects must have the same owner, and the owner must belong to the database (xref) or document that is calling this method.

`NewOwner`   The new owner for the copied objects. If `nil`, the objects will be created with the same owner as the objects in the Objects list.

Returns

A list of newly created duplicate objects. Only primary objects are returned.

Remarks

This method may be used to copy drawing (and non-drawing) objects between spaces, block definitions, drawing, and xrefs.

CopyProfile Method

Copies the specified profile.

`(apa-CopyProfile` *Object OldName NewName*`)`

Arguments

`Object`   The PreferencesProfiles collection to which this method applies.

`OldName`   String, specifying the name of the profile to be copied.

`NewName`   String, specifying the name of the profile to be created.

Returns

T if successful, `nil` otherwise.

Delete Method

Deletes a specified object.

`(apa-Delete` *Object*`)`

Arguments

`Object`   The object (or list of objects) to be deleted.

Returns

`T`

Remarks

When you delete an object from a collection, all remaining items in the collection are reassigned a new index based on the current count. Therefore, while iterating though collections by index, be sure to do so with decreasing index numbers.

Attempting to delete a collection will result in an error.

DeleteConfiguration Method

Deletes a named viewport configuration.

(apa-DeleteConfiguration *Object Name*)

Arguments

Object   The Viewports collection object to which this method applies.

Name   String, specifying the viewport configuration to delete.

DeleteFitPoint Method

Deletes the fit point of a spline at a given index.

(apa-DeleteFitPoint *Object Index*)

Arguments

Object   The Spline object to which this method applies.

Index   Integer specifying the index of fit point to be deleted, starting at 0.

Returns

T if successful, nil otherwise.

DeleteProfile Method

Deletes the specified profile.

(apa-DeleteProfile *Object Name*)

Arguments

Object   The PreferencesProfiles object to which this method applies.

Name   String, specifying the name of the profile to be deleted.

Returns

T if successful, nil otherwise.

Detach Method

Detaches an external reference (xref) from a drawing.

(apa-Detach *Object*)

Arguments

Object   The Block object (xref) to detach.

Returns

T

Display Method

Turns the display control of the PViewport object on or off.

(apa-Display *Object Status*)

Arguments

Object   The PViewport object to be turned on or off.

Status   non-nil   Turns display control on.
                  nil   Turns display control off.

Returns

T

DisplayPlotPreview Method

Displays the Plot Preview dialog box with the specified partial or full view preview.

(apa-DisplayPlotPreview *Object FullPreview*)

Arguments

Object	The Plot object to which this method applies.
FullPreview	non-nil to display a full preview. nil to display a partial preview.

ElevateOrder Method

Elevates the order of the spline to the given order.

(apa-ElevateOrder *Object Order*)

Arguments

Object	Spline.
Order	Number. A positive value greater than the current order. The maximum order is 26.

Returns

T if successful, nil otherwise.

Erase Method

Erases all the objects in a selection set.

(apa-Erase *Object*)

Arguments

Object   The SelectionSet object whose members are to be erased.

Eval Method

Evaluates an expression in VBA.

(apa-Eval *Object Expression*)

Arguments

Object	Application The object or objects this method applies to.
Expression	String, specifying the expression to be evaluated.

This method allows AutoLISP to execute a line of VBA code in the context of the current project without creating modules and functions.

Returns

The value of the Expression as a string.

Evaluate Method

Evaluates the given hatch or leader.

(apa-Evaluate *Object*)

Arguments

Object   The Hatch or Leader object to be evaluated.

Returns

T

Explode Method

Explodes the compound object into sub-entities.

(apa-explode *Object*)

Arguments

Object   3DPoly, BlockRef, LightweightPolyline, MInsertBlock, PolygonMesh, Polyline, or Region object (or list of objects) to be exploded.

Returns

A list of the objects created by the explosion.

Export Method

Exports the AutoCAD drawing to a WMF, SAT, EPS, or DXF format.

`(apa-export Object Filename SelectionSet)`

Arguments

Object	The Document object to be exported.
FileName	String, specifying the name and extension for the exported file.
SelectionSet	SelectionSet object. For WMF and SAT formats, the selection set specifies the objects to be exported. For EPS, DWF, and DXF the selection set is ignored and the entire drawing is exported.

Returns

T if successful, nil otherwise.

ExportProfile Method

Exports a profile to a file.

`(apa-ExportProfile Object ProfileName FileName)`

Arguments

Object	The PreferencesProfiles object to which this method applies.
ProfileName	String, specifying the name of the profile to export.
FileName	String, specifying the name of the file to which to export the profile. Must have an extension of *.arg*.

Returns

T if successful, nil otherwise.

Float Method

Floats the toolbar.

`(apa-get-Float Object Top Left Rows)`

Arguments

Object	The Toolbar object to which this property applies.
Top	A number specifying the pixel location for the top edge of the toolbar.
Left	A number specifying the pixel location for the left edge of the toolbar.
Rows	A number specifying the number of rows for the floating toolbar.

Returns

T

GetAllProfileNames Method

Returns all available profiles for the system.

`(apa-GetAllProfileNames Object)`

Arguments

Object	The PreferencesProfiles object to be queried.

Returns

A list of strings specifying all available profiles for the system.

GetAttributes Method

Returns the non-constant attributes attached to a block reference.

(apa-GetAttributes *Object*)

Arguments

Object   The BlockRef object whose attributes are to be returned.

Returns

A list of AttributeRef objects attached to the block.

Remarks

To find the constant attributes of block reference, use the **apa-GetConstantAttributes** method.

GetBitmaps Method

Returns the large and small bitmaps used as icons for the toolbar item.

(apa-GetBitmaps *Object* '*SmallIconName* '*LargeIconName*)

Arguments

Object   The ToolbarItem to which this method applies.

Returns

SmallIconName   String set to the filename for the small icon.

LargeIconName   String set to the filename for the large icon.

Remarks

SmallIconName and *LargeIconName* must be quoted symbols.

GetBoundingBox Method

Returns the opposite corners of a box enclosing the specified object.

(apa-GetBoundingBox *Object* '*MinPoint* '*MaxPoint*)

Arguments

Object   The drawing object to which this method applies.

Returns

The list (*MinPoint MaxPoint*).

MinPoint   3D point set to the minimum corner of the bounding box, in the WCS.

MaxPoint   3D point set to the maximum corner of the bounding box, in the WCS.

Remarks

MinPoint and *MaxPoint* must be quoted symbols.

The corners are returned in WCS coordinates with the box edges parallel to the WCS *X*, *Y*, and *Z*-axes.

GetBulge Method

Returns the bulge value at a given index of a polyline.

(apa-GetBulge *Object* *Index*)

Arguments

Object   The LightweightPolyline or Polyline object to which this method applies.

Index   Non-negative integer specifying index of the vertex you wish to query.

Returns

Real number specifying the bulge value at the specified index, or nil if not successful.

Remarks

The bulge is the tangent of ¼ the included angle of the following arc segment.

GetCanonicalMediaNames Method

Returns all available plot device names for the system.

(apa-GetPlotDeviceNames *Object*)

Arguments

Object   The Layout or PlotConfiguration object to which this method applies.

Returns

A list of strings specifying all available media names for the specified plot device.

Remarks

You should call **apa-RefreshPlotDeviceInfo** before the first time you use this method. Once you have called **apa-RefreshPlotDeviceInfo**, you don't need to call it again unless your plot device information changes during the session.

GetConstantAttributes Method

Returns the constant attributes in a block or external reference.

(apa-GetConstantAttributes *Object*)

Arguments

Object   The BlockRef or ExternalReference object whose constant attributes are to be returned.

Returns

A list of constant Attribute definitions contained in the block definition.

Remarks

To find the non-constant attributes of block reference, use the **apa-GetAttributes** method.

GetControlPoint Method

Returns the coordinates of the control point of a spline at a given index.

(apa-GetControlPoint *Object Index*)

Arguments

Object   The Spline object to which this method applies.

Index   Non-negative integer specifying index of the control point you wish to query.

Returns

The coordinates of the control point at the given index location as a 3D point, in the WCS, or nil if not successful.

GetCustomScale Method

Returns the custom scale for a layout or plot configuration.

(apa-GetCustomScale *Object 'Numerator 'Denominator*)

Arguments

Object   The Layout or PlotConfiguration object to which this method applies.

Returns

The list (*Numerator Denominator*).

Numerator   Real number set to the numerator of the scale ratio.

Denominator   Real number set to the denominator of the scale ratio.

Remarks

Numerator and *Denominator* must be quoted symbols.

The *Numerator* is the 'plotted units' part of the scale.

The *Denominator* is the 'drawing units' part of the scale.

GetFitPoint Method

Returns the coordinates of the fit point of a spline at a given index.

(apa-GetFitPoint *Object Index*)

Arguments

Object	The Spline object to which this method applies.
Index	Non-negative integer specifying index of the fit point you wish to query.

Returns

The coordinates of the fit point at the given index location as a 3D point, in the WCS, or nil if not successful.

GetFont Method

Returns the definition data of the font for the TextStyle.

(apa-GetFont *Object* 'Typeface 'Bold 'Italic 'CharSet 'PitchAndFamily)

Arguments

Object	The TextStyle object to which this method applies.

Returns

The list *(Typeface Bold Italic CharSet PitchAndFamily)*

Typeface	A string set to the typeface (font name) of the TextStyle you queried.
Bold	Set to T if the TextStyle is bold, nil otherwise.
Italic	Set to T if the TextStyle is italic, nil otherwise.
CharSet	Set to an integer representing the character set for the font. (See the available values in the Remarks section.)
PitchAndFamily	Set to an integer representing the pitch and family definitions for the font. (See the available values in the Remarks section.)

Remarks

The *CharSet* parameter specifies the character set for the font.

apANSI_CHARSET	0
apDEFAULT_CHARSET	1
apSYMBOL_CHARSET	2
apSHIFTJIS_CHARSET	128
apOEM_CHARSET	255

The *PitchAndFamily* parameter specifies the pitch and family values for the font. The value is determined by a combination of two separate settings.

By choosing a setting from each of the two categories and using the **logior** function to combine them, you achieve the *PitchAndFamily* value.

Pitch Values

apDEFAULT_PITCH	0
apFIXED_PITCH	1
apVARIABLE_PITCH	2

Family Values

apFF_DONTCARE	0	Don't care or don't know.
apFF_ROMAN	16	Variable stroke width, serifed.
apFF_SWISS	32	Variable stroke width, sans-serifed.
apFF_MODERN	48	Constant stroke width, serifed or sans-serifed.

| apFF_SCRIPT | 64 | Cursive, etc. |
| apFF_DECORATIVE | 80 | Old English, etc. |

GetGridSpacing Method

Returns the grid spacing for the viewport.

(apa-GetGridSpacing *Object* '*XSpacing* '*YSpacing*)

Arguments

Object   The Viewport or PViewport object to be queried.

Returns

The list *(XSpacing YSpacing)*

XSpacing   Real number set to XSpacing of the grid.

YSpacing   Real number set to YSpacing of the grid.

Remarks

XSpacing and *YSpacing* must be quoted symbols.

GetInvisibleEdge Method

Returns the invisibility setting for an edge of a 3DFace object at a given index.

(apa-GetInvisibleEdge *Object* *Index*)

Arguments

Object   The 3DFace object to which this method applies.

Index   Integer, in the range 0-3, specifying the index of the edge to be queried.

Returns

T if the edge is invisible, nil if it's not.

GetLoopAt Method

Returns the hatch loop at a given index.

(apa-GetLoopAt *Object* *Index*)

Arguments

Object   The Hatch object to which this method applies.

Index   Non-negative integer index of the loop to be returned.

Returns

A list of objects comprising the loop. Loop 0 is the outer loop.

GetPaperMargins Method

Returns the margins for the layout or plot configuration.

(apa-GetPaperMargins *Object* '*LowerLeft* '*UpperRight*)

Arguments

Object   The Layout or PlotConfiguration object to which this method applies.

Returns

The list *(LowerLeft UpperRight)*.

LowerLeft   2D point set to lower left margin.

UpperRight   2D point set to upper right margin.

Remarks

LowerLeft and *UpperRight* must be quoted symbols.

GetPaperSize Method

Returns the width and height of the configured paper.

`(apa-GetPaperSize Object 'Width 'Height)`

Arguments

`Object`   The Layout or PlotConfiguration object to which this method applies.

Returns

The list `(Width Height)`.

`Width`   Real number set to the width of the paper.

`Height`   Real number set to the height of the paper.

Remarks

`Width` and `Height` must be quoted symbols.

GetPlotDeviceNames Method

Returns all available plot device names for the system.

`(apa-GetPlotDeviceNames Object)`

Arguments

`Object`   The Layout or PlotConfiguration object to which this method applies.

Returns

A list of strings specifying all available plot devices for the system.

Remarks

You should call **apa-RefreshPlotDeviceInfo** before the first time you use this method. Once you have called **apa-RefreshPlotDeviceInfo**, you don't need to call it again unless your plot device information changes during the session.

GetPlotStyleTableNames Method

Returns all available plot style table names for the system.

`(apa-GetPlotStyleTableNames Object)`

Arguments

`Object`   The Layout or PlotConfiguration object to be queried.

Returns

A list of strings specifying all available plot style table names for the system.

Remarks

You should call **apa-RefreshPlotDeviceInfo** before the first time you use this method. Once you have called **apa-RefreshPlotDeviceInfo**, you don't need to call it again unless your plot device information changes during the session.

GetProjectFilePath Method

Returns the Project File Path for a given project.

`(apa-GetProjectFilePath Object ProjectName)`

Arguments

`Object`   The PreferencesFiles object to which this method applies.

`ProjectName`   String, specifying the name of the project whose file path we wish to retrieve.

Returns

String containing the Project File Path, or `nil` if not successful.

GetSnapSpacing Method

Returns the snap spacing for the viewport.

(apa-GetSnapSpacing *'XSpacing 'YSpacing*)

Arguments

Object   The Viewport or PViewport object to which this method applies.

Returns

The list *(XSpacing YSpacing)*

XSpacing   Real number set to XSpacing of the snap.

YSpacing   Real number set to YSpacing of the snap.

Remarks

XSpacing and *YSpacing* must be quoted symbols.

GetUCSMatrix Method

Returns the transformation matrix for the specified UCS.

(apa-GetUCSMatrix *Object*)

Arguments

Object   The UCS object to which this method applies.

Returns

Transformation Matrix (4×4 list of Reals) for the specified UCS.

Remarks

Use **apa-transformby** and the matrix returned by this method to transform an entity to the given UCS.

GetVariable Method

Returns the current setting of an AutoCAD system variable.

(apa-GetVariable *Object Name*)

Arguments

Object   The Document object to be queried.

Name   String representing the name of the variable to be returned.

Returns

The current setting of the system variable.

Remarks

If you're querying the current drawing, it's probably faster to use the **getvar** function. If, however, you wish to query *another* drawing, this is *the* way to go.

GetWeight Method

Returns the weight of the control point of a spline at a given index.

(apa-GetWeight *Object Index*)

Arguments

Object   The Spline object to be queried.

Index   Non-negative integer index of the control point to query.

Returns

The weight of the specified control point as a real, or **nil** if unsuccessful

GetWidth Method

Returns the starting and ending widths for a segment of a polyline.

(apa-GetWidth *Object Index* 'StartWidth 'EndWidth)

Arguments

Object	The LightweightPolyline or Polyline object to be queried.
Index	Non-negative integer index of the segment to query.

Returns

The list (*StartWidth EndWidth*).

StartWidth	Real number set to the starting width of the specified segment.
EndWidth	Real number set to the ending width of the specified segment.

Remarks

StartWidth and *EndWidth* must be quoted symbols.

GetWindowToPlot

Returns the coordinates that define the portion of the layout to plot.

(apa-GetWindowToPlot *Object* 'LowerLeft 'UpperRight)

Arguments

Object	The Layout or PlotConfiguration object to which this method applies.

Returns

The list *(LowerLeft UpperRight)*

LowerLeft	2D point set to lower left corner.
UpperRight	2D point set to upper right corner.

Remarks

LowerLeft and *UpperRight* must be quoted symbols.

GetXData Method

Returns the extended data (XData) associated with an object.

(apa-GetXData *Object AppName*)

Arguments

Object	The object whose XData you wish to retrieve.
AppName	String, specifying the application whose XData should be retrieved. An empty string (or nil) returns all the application XData for the object.

Returns

List containing the XData for the specified object.

HandleToObject Method

Returns the object that corresponds to the given handle.

(apa-HandleToObject *Handle*)

Arguments

Object	The Document object to which this method applies.
Handle	String, specifying the handle of the object to retrieve.

Returns

The object corresponding to the handle, or nil if not successful.

Remarks

This method can return objects from only the current document.

Highlight Method

Sets the highlight status for the given object, or for all objects in a given selection set or group.

(apa-Highlight *Object Highlight*)

Arguments

Object	The Drawing, SelectionSet or Group object (or list of objects) to be highlighted or un-highlighted.
Highlight	non-nil highlights the specified object. nil un-highlights the object.

Returns

T

Remarks

Once the highlight flag for an object has been changed, a call to the Update method may be required to view the change.

Import Method

Imports a drawing file in SAT, EPS, DXF, or WMF format.

(apa-Import *Object FileName InsertionPoint ScaleFactor*)

Arguments

Object	The drawing object to which this method applies.
FileName	String, specifying the file to import.
InsertionPoint	Point at which the imported file is to be inserted, in the WCS.
ScaleFactor	Positive number specifying the scale factor for the import.

Returns

If successful, a BlockReference object is returned when importing a WMF file, T otherwise.

If unsuccessful, returns nil.

Remarks

The object will be imported into the ModelSpace collection of the drawing object.

Searches the AutoCAD Search Path.

ImportProfile Method

Imports a profile from an external file.

(apa-ImportProfile *Object ProfileName FileName IncludePath*)

Arguments

Object	The PreferencesProfiles collection object.
ProfileName	String, specifying the description of the profile to import.
FileName	String, specifying the name of the *.arg* file to be imported.
IncludePath	non-nil to preserve the path information in the *.arg* file. nil to not preserve it.

Returns

T if successful, nil otherwise.

InsertBlock Method

Inserts a drawing file or a named block parallel to the XY plane of the UCS.

```
(apa-InsertBlock Object Name InsertionPoint XScale YScale ZScale
        RotationAngle)
```

Arguments

Object	ModelSpace, PaperSpace, or Block collection object in which to add the newly created object.
Name	String, specifying the name of the AutoCAD drawing file or the name of the block to insert. If it is a file name, include any path information necessary for AutoCAD to find the file and the .dwg extension.
InsertionPoint	Point specifying the insertion point of the block, in the WCS.
XScale	Non-zero number specifying the X scale factor of the block.
YScale	Non-zero number specifying the Y scale factor of the block.
ZScale	Non-zero number specifying the Z scale factor of the block.
RotationAngle	Number specifying the rotation angle in radians for the block.

Returns

The newly created BlockRef object, or `nil` if not successful.

Remarks

Searches the AutoCAD Search Path.

InsertLoopAt Method

Inserts a loop at a given index of a hatch.

```
(apa-InsertLoopAt Object Index LoopType Loop)
```

Arguments

Object	The Hatch object to which the loop should be inserted.
Index	Non-negative integer specifying where to insert the loop.
LoopType	One of the following constants:
	acHatchLoopTypeDefault
	acHatchLoopTypeExternal
	acHatchLoopTypePolyline
	acHatchLoopTypeDerived
	acHatchLoopTypeTextbox
Loop	List of Line, Polyline, Circle, Ellipse, Spline, and Region objects comprising the loop.

IntersectWith Method

Returns the points where one object intersects another object in the drawing.

```
(apa-IntersectWith ThisObject OtherObject ExtendOption)
```

Arguments

ThisObject	First Drawing object to be checked for intersection.
OtherObject	Second Drawing object to be checked for intersection.
ExtendOption	One of the following constants:

acExtendNone	Does not extend either object.
acExtendThisEntity	Extends ThisObject.
acExtendOtherEntity	Extends OtherObject.
acExtendBoth	Extends both objects.

Returns

List of 3D points, in the WCS, where one object intersects the other.

Item Method

Returns the member object at a given index in a collection, group, or selection set.

(apa-Item *Object Index*)

Arguments

Object	The Collection, Group, or SelectionSet object to query.
Index	The index location in the collection for the member item to query. Either: A non-negative integer specifying the index number of the member item, OR A string, specifying the name of the member item.

Returns

If successful, the object at the given index location in the collection, group, or selectionset.

If not successful, nil.

Load Method – Linetype

Loads the definition of a linetype from a library (LIN) file.

(apa-LoadLinetype *Object LinetypeName FileName*)

Arguments

Object	The Linetypes collection object to which this method applies.
LinetypeName	String, specifying the name of the linetype to load. Wildcards are acceptable.
FileName	String, specifying name of the linetype file to load.

Returns

T if successful, nil otherwise.

Remarks

You cannot load a linetype that is already loaded.

Searches the AutoCAD Search Path.

Load Method – Menu

Loads a menu group from a menu file.

(apa-LoadMenu *Object MenuFileName BaseMenu*)

Arguments

Object	The MenuGroups collection object to which this method applies. are acceptable.
MenuFileName	String, specifying name of the menu file to load. Should have an extension of *.mnc*, *.mns*, or *.mnu*.
BaseMenu	non-nil to load the menu group as a base menu, as with the MENU command. nil to load the menu group as a partial menu, as with the MENULOAD command.

Returns

The newly loaded MenuGroup object, or nil if not successful.

Remarks

You cannot load a MenuGroup that is already loaded.

Searches the AutoCAD Search Path.

LoadShapeFile Method

Loads a shape (SHX) file.

```
(apa-LoadShapeFile Object FileName)
```
Object   The Document object to which this method applies.

FileName   String, specifying the name of the shape (*.shx*) file to load.

Returns

T if successful, nil otherwise.

Remarks

You cannot load a shape file that is already loaded.

Searches the AutoCAD Search Path.

Mirror Method

Creates a mirror image copy of an object, parallel to the XY plane of the UCS.

```
(apa-Mirror Object Point1 Point2)
```

Arguments

Object   The Drawing object (or list of objects) to be mirrored.

Point1   Point specifying the first point of the mirror line, in the WCS.

Point2   Point specifying the second point of the mirror line, in the WCS.

Returns

The newly created mirrored object (or list of objects).

Mirror3D Method

Creates a mirror image of the given object about a plane.

```
(apa-Mirror3D Object Point1 Point2 Point3)
```

Arguments

Object   The Drawing object (or list of objects) to be mirrored.

Point1   Point specifying the first point of the mirror plane, in the WCS.

Point2   Point specifying the second point of the mirror plane, in the WCS.

Point3   Point specifying the third point of the mirror plane, in the WCS.

Returns

The newly created mirrored object (or list of objects).

Move Method

Moves an object along a vector.

```
(apa-Move Object Point1 Point2)
```

Arguments

Object   The Drawing object (or list of objects) to be moved.

Point1   Point specifying the first point of the vector, in the WCS.

Point2   Point specifying the second point of the vector, in the WCS.

Returns

T

ObjectIDToObject Method

Returns the object that corresponds to the given object ID.

(apa-ObjectIDToObject *Object ObjectID*)

Arguments

Object	The Document object to which this method applies.
ObjectID	Integer specifying the object ID of the object to return.

Returns

The object that corresponds to the given object ID.

Remarks

This method can return objects from only the current document.

Offset Method

Creates a new object at a specified offset distance from an existing object.

(apa-Offset *Object Distance*)

Arguments

Object	The Arc, Circle, Ellipse, Line, LightweightPolyline, Polyline, Spline, or XLine object (or list of objects) to be offset.
Distance	Non-zero number specifying the distance of offset the object.
	If the distance is negative, it's interpreted as specifying an offset to create a 'smaller' curve. If 'smaller' has no meaning, the object is offset in the direction of smaller *X*, *Y*, and *Z* WCS coordinates.

Returns

A list of the newly created objects.

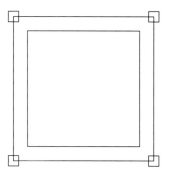

Open Method

Opens an existing drawing file (DWG).

(apa-Open *Object FileName ReadOnly*)

Arguments

Object	The Documents collection to which this method applies.
FileName	String, specifying the path or URL of the .dwg file to open.
ReadOnly	non-nil to open the drawing in ReadOnly mode. nil to open for ReadWrite.

Returns

The newly created document object, or nil if not successful.

Remarks

This method opens AutoCAD drawing files (DWG) only. Use the Import method to open SAT, EPS, DXF, or WMF format files.

Do not use this method while in SDI mode.

The AutoCAD search path is *not* searched.

Use the **apa-active** function to set the active document.

PlotToDevice

Plots a layout to a device.

(apa-PlotToDevice *Object PlotConfig*)

Arguments

Object	The Plot object to which this method applies.
PlotConfig	The filename of the PC3 file to use instead of the current configuration, or nil.
	If the file isn't found, AutoCAD will search the plotter configuration path for the file.

Returns

T if the plot was successful, nil otherwise.

Remarks

The plot must be initiated from the active drawing.

The plot device is specified using the ConfigName property for the layout or plot configuration.

Should the specified PC3 file or the current plot configuration specify plot to file information, the plot may be sent to a file instead of the device.

PlotToFile Method

Plots a layout to the specified file.

(apa-PlotToFile *Object PlotFile PlotConfig*)

Arguments

Object	The Plot object to which this method applies.
PlotFile	String, specifying the name of the plot file. When multiple plots are produced, the file name will be generated from the drawing and layout names.
PlotConfig	nil, or a string, specifying the filename of the PC3 file to use instead of the current configuration. If the file isn't found, AutoCAD will search the plotter configuration path for the file.

Returns

T if the plot was successful, nil otherwise.

Remarks

The plot must be initiated from the active drawing.

The plot device is specified using the ConfigName property for the layout or plot configuration.

Should the specified PC3 file or the current plot configuration specify plot to file information, the plot may be sent to a file instead of the device.

PurgeAll Method

Purges unused layers, blocks, etc.

(apa-PurgeAll *Object*)

Arguments

Object	The Document object to be purged.

Returns

T

Remarks

Equivalent to Purge All from command line.

PurgeFitData Method

Purges the fit data of a spline.

(apa-PurgeFitData *Object*)

Arguments

Object   The Spline object from which to remove the fit points

Returns

T

Regen Method

Regenerates the entire drawing.

(apa-Regen *Object AllViewports*)

Arguments

Object   The Document object to be regenerated.

AllViewports   non-nil to regenerate all viewports in the document, nil to regenerate only the active
                         viewport.

Returns

T

RefreshPlotDeviceInfo Method

Updates the plot, canonical media, and plot style table information to reflect the current system state.

(apa-RefreshPlotDeviceInfo *Object*)

Arguments

Object   The Layout or PlotCofiguration object to which this method applies.

Remarks

It is recommended that you refresh your plot device information before you first use the
ap-GetCanonicalMediaNames, **ap-GetPlotDeviceNames**, or **ap-GetPlotStyleTableNames** functions
for a given AutoCAD session. After that, you need only refresh the information if some part of the device setup
changes during the course of the session.

Reload Method

Reloads the specified external reference.

(apa-Reload *Object*)

Arguments

Object   The ExternalReference object to be reloaded.

Returns

T

Rename Method

Renames an item in the specified dictionary.

(apa-Rename *Object OldName NewName*)

Arguments

Object	The Dictionary object to which this method applies.
OldName	String, specifying the name (keyword) of the item to be renamed.
NewName	String, specifying the new name for the item.

RenameProfile Method

Renames the specified profile.

`(apa-RenameProfile `*Object OldName NewName*`)`

Arguments

Object	The PreferencesProfiles object to which this method applies.
OldName	String, specifying the name of the profile to be renamed.
NewName	String, specifying the new name for the profile.

Returns

T if successful, `nil` otherwise.

ResetProfile Method

Resets the values in the specified profile to the system default settings.

`(apa-ResetProfile `*Object Name*`)`

Arguments

Object	The PreferencesProfiles object to which this method applies.
Name	String, specifying the name of the profile to be reset.

Returns

T if successful, `nil` otherwise.

Remarks

The specified profile need not be the current active profile.

Reverse Method

Reverses the direction of a spline.

`(apa-Reverse `*Object*`)`

Arguments

Object	The Spline object to be reversed.

Returns

T

Rotate Method

Rotates an object around a base point, parallel to the XY plane of the UCS.

`(apa-Rotate `*Object BasePoint RotationAngle*`)`

Arguments

Object	The Drawing object (or list of objects) to be rotated.
BasePoint	Point specifying a point on the axis of rotation, in the WCS. The axis of rotation is parallel to the Z-axis of the UCS.
RotationAngle	Number specifying the rotation angle in radians for the object.

Returns

T

Rotate3D Method

Rotates an object around a specified 3D axis.

```
(apa-Rotate3D Object Point1 Point2 RotationAngle)
```

Arguments

Object	The Drawing object to be rotated.
Point1	Point specifying the first point of rotation axis.
Point2	Point specifying the second point of the rotation axis.
RotationAngle	Number specifying the rotation angle in radians for the object.

Save Method

Saves the specified document.

```
(apa-SaveDocument Object)
```

Arguments

Object	The Document object to be saved.

Returns

T if successful, nil otherwise.

Save Method – MenuGroup

Saves the specified menugroup.

```
(apa-SaveMenuGroup Object FileType)
```

Arguments

Object	The MenuGroup object to be saved.
FileType	One of the following:

acMenuFileCompiled: A compiled menu file (*.mnc).
acMenuFileSource: A source menu file (*.mns).

Returns

T if successful, nil otherwise.

SaveAs Method

Saves the document or menu group to a specified file.

```
(apa-SaveAs Object FileName FileType)
```

Object	The Document or MenuGroup object to be saved.
FileName	String, specifying the name of the file.
	The specified document or menu group takes on the new name.
FileType	For Drawing files:

acR12_DXF:	AutoCAD R12/LT2 DXF (*.dxf)
acR13_DWG:	AutoCAD R13/LT95 DWG (*.dwg)
acR13_DXF:	AutoCAD R13/LT95 DXF (*.dxf)
acR14_DWG:	AutoCAD R14/LT97 DWG (*.dwg)
acR14_DXF:	AutoCAD R14/LT97 DXF (*.dxf)
acR15_DWG:	AutoCAD 2000 DWG (*.dwg)
acR15_DXF:	AutoCAD 2000 DXF (*.dxf)
acR15_Template:	AutoCAD 2000 Drawing Template File (*.dwt)
acNative :	A synonym for the latest drawing release.

For Menu Files:

acMenuFileCompiled:	A compiled menu file (*.mnc).
acMenuFileSource:	A source menu file (*.mns).

Returns

T if successful, nil otherwise.

Remarks

The active document or menu group takes on the new name.

Use the **apa-Export** function to create a DXF file without changing the name of the document.

ScaleEntity Method

Uniformly scales an object.

```
(apa-Scale Object BasePoint ScaleFactor)
```

Arguments

Object	The Drawing object (or list of objects) to be scaled.
BasePoint	Point specifying the base point, in the WCS.
ScaleFactor	Positive number specifying the factor by which to scale the object.

Returns

T

Select Method

Adds the selected objects to a selection set.

```
(apa-Select Object Mode Point1 Point2 Filter)
```

Arguments

Object	The SelectionSet object to which this property applies.
Mode	One of the following:
	acSelectionSetWindow
	acSelectionSetCrossing
	acSelectionSetPrevious
	acSelectionSetLast
	acSelectionSetAll
Point1	Point specifying the first point selection window.
Point2	Point specifying the second point of the selection window.
Filter	A selection set filter list.

Returns

T if at least one object is in the selection set, nil otherwise.

Remarks

With the exception of **acSelectionSetAll** mode, objects must be on screen to be selected.

SelectAtPoint Method

Adds an object at the specified point to the selection set.

```
(apa-SelectAtPoint Object Point1 Filter)
```

Arguments

Object	The SelectionSet object to which this property applies.
Point1	Point specifying the point for the selection.
Filter	A selection set filter list.

Returns

T if at least one object is in the selection set, nil otherwise.

Remarks

Objects must be on screen to be selected.

SelectByPolygon Method

Adds an object at the specified point to the selection set.

(apa-SelectByPolygon *Object Mode Points Filter*)

Arguments

Object	The SelectionSet object to which this property applies.
Mode	One of the following:
	acSelectionSetFence
	acSelectionSetWindowPolygon
	acSelectionSetCrossingPolygon
Points	A list of points.
Filter	A selection set filter list.

Returns

T if at least one object is in the selection set, nil otherwise.

Remarks

Objects must be on screen to be selected.

SelectOnScreen Method

Prompts and allows the user to select objects on screen.

(apa-SelectOnScreen *Object Filter*)

Arguments

Object	The SelectionSet object to which this property applies.
Filter	A selection set filter list.

Returns

T if at least one object is in the selection set, nil otherwise.

SectionSolid Method

Creates a region that represents the intersection of a solid and a plane.

(apa-SectionSolid *Object Point1 Point2 Point3*)

Arguments

Object	The Solid object (or list of objects) to be sectioned.
Point1	Point specifying the first point of the section plane, in the WCS.
Point2	Point specifying the second point of the section plane, in the WCS.
Point3	Point specifying the third point of the section plane, in the WCS.

Returns

The newly created Region object (or list of objects).

SendCommand Method

Sends a text string to the AutoCAD interface in the context of the active document.

(apa-SendCommand *Object Command*)

Arguments

Object	The document object to receive the command.
Command	A string, specifying the command to be sent to the document. Be sure to include a \n or space at the end of the command if you wish it to be executed.

Returns

T

Remarks

If the specified document is not the active document, it is made the active document, and control does not pass back to the calling routine until its document becomes the active document.

SetBulge Method

Sets the bulge value at a given index of a polyline.

(apa-SetBulge *Object Index Bulge*)

Arguments

Object	The LightweightPolyline or Polyline object to which this method applies.
Index	Non-negative integer specifying index of the vertex you wish to query.
Bulge	Number specifying the bulge value at the specified index.

Returns

T

Remarks

The bulge is the tangent of ¼ the included angle of the following arc segment. It's also twice the ratio of the height of the arc segment to its chord.

SetControlPoint Method

Sets the coordinates of the control point of a spline at a given index.

(apa-SetControlPoint *Object Index Point*)

Arguments

Object	The Spline object to which this method applies.
Index	Non-negative integer specifying index of the control point you wish to query.
Point	Point specifying coordinates of the control point at the given index location, in the WCS.

Returns

T

SetCustomScale Method

Sets the custom scale for a layout or plot configuration.

(apa-SetCustomScale *Object Numerator Denominator*)

Arguments

Object	The Layout or PlotConfiguration object to which this method applies.
Numerator	Number specifying numerator of the scale ratio.
Denominator	Number specifying the denominator of the scale ratio.

Returns

T if successful, nil otherwise.

Remarks

The *Numerator* is the 'plotted units' part of the scale.

The *Denominator* is the 'drawing units' part of the scale.

SetFitPoint Method

Sets the coordinates of the fit point of a spline at a given index.

(apa-SetFitPoint *Object Index Point*)

Arguments

Object	The Spline object to which this method applies.
Index	Non-negative integer specifying index of the fit point you wish to query.
Point	Point specifying the coordinates of the fit point at the given index location, in the WCS.

Returns

T

SetInvisibleEdge Method

Sets the invisibility setting for an edge of a 3DFace object at a given index.

`(apa-SetInvisibleEdge Object Index Invisible)`

Arguments

Object	The 3DFace object to which this method applies.
Index	Integer, in the range 0-3, specifying the index of the edge to be queried.
Invisible	non-`nil` to set the edge invisible, `nil` to set it visible.

Returns

T

SetLayoutsToPlot

Specifies the layout or layouts to plot.

`(ap-SetLayoutsToPlot Object LayoutList)`

Arguments

Object	The plot objects to which this method applies.
LayoutList	A list of layout names representing the layouts to plot.

Returns

T

Remarks

To specify any layout other than the active layout, you must call SetLayoutsToPlot method before each plot.

SetPattern Method

Sets the pattern name and pattern type for the hatch.

`(apa-SetPattern Object PatternType PatternName)`

Arguments

Object	The Hatch object to which this method applies.

PatternType	One of the following values:
	`acHatchPatternTypePredefined`
	`acHatchPatternTypeUserDefined`
	`acHatchPatternTypeCustomDefined`

PatternName	String, specifying the name of the hatch pattern.

Returns

T

SetProjectFilePath Method

Sets the Project File Path for a given project.

`(apa-SetProjectFilePath Object ProjectName ProjectPath)`

Returns

T

Arguments

Object	The PreferencesFiles object to which this method applies.
ProjectName	String, specifying the name of the project whose file path we wish to set.
ProjectPath	String, specifying the Project File Path.

SetVariable Method

Sets the current setting of an AutoCAD system variable.

(apa-SetVariable *Object Name NewValue*)

Arguments

Object	The Document object to be queried.
Name	String representing the name of the variable to be returned.
NewValue	New value for specified variable.

Returns

Prior value of system variable.

Remarks

If you're setting a variable in the current drawing, it's probably faster to use the **setvar** function. If, however, you wish to set a variable in *another* drawing, this is *the* way to go.

SetView Method

Sets the view in a viewport to a saved view in the Views Collection object.

(apa-SetView *Object View*)

Arguments

Object	The Viewport object to which this method applies.
View	The View object specifying the view to set.

Returns

T

SetWeight Method

Sets the weight of the control point of a spline at a given index.

(apa-SetWeight *Object Index Weight*)

Arguments

Object	The Spline object to be modified.
Index	Non-negative integer index of the control point to set.
Weight	Number specifying the new weight of the control point.

Returns

T

SetWidth Method

Sets the starting and ending width for a segment of a polyline.

(apa-SetWidth *Object Index StartWidth EndWidth*)

Arguments

Object	The LightweightPolyline or Polyline object to be queried.
Index	Non-negative integer index of the segment to query.
StartWidth	Non-negative number specifying the starting width of the specified segment.
EndWidth	Non-negative number specifying the ending width of the specified segment.

Returns

T

SetXData Method

Sets the extended data (XData) associated with an object.

(apa-SetXData *Object AppName XData*)

Arguments

Object	The object whose XData you wish to retrieve.
AppName	String, specifying the application whose XData should be set.
XData	List containing the XData for the specified object.

Returns

T

SetXRecordData Method

Sets the extended record data (XRecordData) associated with a dictionary.

(apa-SetXRecordData *Object XRecordData*)

Arguments

Object	The Dictionary or XRecord object whose XRecordData you wish to retrieve.
XRecordData	List containing the XRecordData for the specified object.

SliceSolid Method

Creates a slice of a 3DSolid object given three points that define the plane.

(apa-SectionSolid *Object Point1 Point2 Point3 Positive*)

Arguments

Object	The Solid object (or list of objects) to be sliced.
Point1	Point specifying the first point of the slicing plane, in the WCS.
Point2	Point specifying the second point of the slicing plane, in the WCS.
Point3	Point specifying the third point of the slicing plane, in the WCS.
Positive	non-nil to return the slice on the positive side of the slicing plane. nil to discard it.

Returns

If Positive is non-nil, the slice on the positive side of the slicing plane as a Solid object (or list of objects).

Remarks

The slice on the negative side of the slicing plane replaces the original solid.

Split Method

Splits a viewport into the given number of views.

(apa-Split *Object NumWins*)

Arguments

Object	The Viewport object to be split.
NumWins	Number specifying how the viewport is to be split. One of the following:

```
acViewport2Horizontal
acViewport2Vertical
acViewport3Left
acViewport3Right
acViewport3Horizontal
acViewport3Vertical
```

```
acViewport3Above
acViewport3Below
acViewport4
```

Returns

T

Remarks

A viewport need not to be active for this method to work.

The viewport must be set active to realize the results of this method.

TransformBy Method

Transforms an object by the specified transformation matrix.

```
(apa-TransformBy Object XformMatrix)
```

Arguments

Object   The Drawing object (or list of objects) to be transformed.

XformMatrix   Transformation Matrix (4×4 list of reals) specifying the transformation to be performed.

Returns

T

Remarks

See *Transformation Matrices*.

Unload Method

Unloads a menu group or external reference.

```
(apa-Unload Object)
```

Arguments

Object   The MenuGroup or External Reference object to be unloaded.

Returns

T

Update Method

Updates the object to the drawing screen.

```
(apa-Update Object)
```

Arguments

Object   The Drawing object to be updated.

Returns

T

Remarks

Because you *might* be making more than one modification to an object, AutoCAD defers updating the display of an object until it is certain you will not be making any more modifications.

AutoCAD's strategy seems to be not to update one object until you start modifying another.

Since each apa-Update takes time, it behooves us to use a *minimum* number of apa-Updates. At most, you should use one on *the very last* object modified before pausing for some user input.

There seems to be no reason to have one at the *end* of a command.

WBlock Method

Creates a drawing file from a selection set.

```
(apa-WBlock Object FileName SelectionSet)
```

Arguments

Object	The Document object containing the selection set.
FileName	String, specifying the FileName of the drawing file (.DWG) to create.
SelectionSet	SelectionSet object specifying the objects to extract.

ZoomAll Method

Zooms the current viewport to display the entire drawing.

(apa-ZoomAll *Object*)

Arguments

Object	The Application object to which this method applies.

Returns

T

ZoomCenter Method

Zooms the current viewport to a specified center point and magnification.

(apa-ZoomCenter *Object Center Height*)

Arguments

Object	The Application object to which this method applies.
Center	Point specifying the new center of the window, in the WCS.
Height	Positive number specifying the height of the viewport in drawing units.

Returns

T

ZoomExtents Method

Zooms the current viewport to the drawing extents.

(apa-ZoomExtents *Object*)

Object	The Application object to which this method applies.

Returns

T

ZoomPickWindow Method

Zooms the current viewport to a window defined by points picked by the user on the screen.

(apa-ZoomPickWindow *Object*)

Arguments

Object	The Application object to which this method applies.

Returns

T

Remarks

The user's canceling the zoom does not create an error condition.

ZoomPrevious Method

Zooms the current viewport to a window defined by points picked by the user on the screen.

(apa-ZoomPrevious *Object*)

Arguments

Object	The Application object to which this method applies.

Returns

T

ZoomScaled Method

Zooms the current viewport by the given scale factor.

(apa-ZoomScaled *Object Scale ScaleType*)

Arguments

Object	The Application object to which this method applies.
Scale	Positive number specifying the zoom factor for the zoom.
ScaleType	One of the following:

    acZoomScaledAbsolute   n
    acZoomScaledRelative   (nX)
    acZoomScaledRelativePSpace (nXP)

Returns

T

Remarks

When using, acZoomScaledAbsolute *ScaleType*, the *Scale* value multiplies the apparent display size of any objects from what it would be if you were zoomed to the limits of the drawing.

ZoomWindow Method

Zooms the current viewport to the area specified by two corners of a rectangle.

(apa-ZoomWindow *Object Point1 Point2*)

Arguments

Object	The Application object to which this method applies.
Point1	Point specifying the first corner of the zoom rectangle, in the WCS.
Point2	Point specifying the second corner of the zoom rectangle, in the WCS.

Returns

T

Artful Programming API ActiveX Property Functions

Active Property

Determines if the document is the active document for the session.

(apa-get-Active *Object*)

Arguments

Object	The Document object to which this property applies.

Returns

T if the specified document is the active document, nil otherwise.

ActiveDimStyle Property

Specifies the active dimension style.

(apa-get-ActiveDimStyle *Object*)
(apa-set-ActiveDimStyle *Object ActiveDimStyle*)

Arguments

Object	The Document object to which this property applies.
ActiveDimStyle	DimStyle Object to become the active dimension style.

Returns

The active dimension style as a DimStyle Object.

Remarks

This style will be applied to all newly created dimensions. To change the style on an existing dimension, use the **ap-put-StyleName** function.

Dimensions created via the AutoCAD user interface are created with the active dimension style plus all document overrides. Dimensions created via ActiveX are created with the active dimension style only.

ActiveDocument Property

Specifies the active document (drawing file).

```
(apa-get-ActiveDocument Object)
(apa-set-ActiveDocument Object ActiveDocument)
```

Arguments

Object	The Application object to which this property applies.
ActiveDocument	The Document Object to become the active dimension style.

Returns

The active document as a Document object.

ActiveLayer Property

Specifies the active layer.

```
(apa-get-ActiveLayer Object)
(apa-set-ActiveLayer Object ActiveLayer)
```

Arguments

Object	The Application object to which this property applies.
ActiveLayer	The Layer object to become the active layer.

Returns

The active layer as a Layer object.

ActiveLayout Property

Specifies the active layout.

```
(apa-get-ActiveLayout Object)
(apa-set-ActiveLayout Object ActiveLayout)
```

Arguments

Object	The Application object to which this property applies.
ActiveLayout	The Layout object to become the active layout.

Returns

The active layout as a Layout object.

ActiveLinetype Property

Specifies the active linetype for the drawing.

```
(apa-get-ActiveLinetype Object)
(apa-set-ActiveLinetype Object ActiveLinetype)
```

Arguments

Object	The Application object to which this property applies.
ActiveLinetype	The Linetype object to become the active linetype.

Returns

The active Linetype as a Linetype object.

ActiveProfile Property

Specifies the name of the active profile.

```
(apa-get-ActiveProfile Object)
(apa-set-ActiveProfile Object Name)
```

Arguments

Object	The PreferencesProfiles object to which this property applies.
Name	A string, specifying the name of the active profile.

Returns

A string.

ActivePViewport Property

Specifies the active paper space viewport for the drawing.

```
(apa-get-ActivePViewport Object)
(apa-set-ActivePViewport Object ActivePViewport)
```

Arguments

Object	The Application object to which this property applies.
ActivePViewport	The PViewport object to become the active paper space viewport.

Returns

The active paper space viewport as a PViewport object.

ActiveSelectionSet Property

Returns the active selection set for the drawing.

```
(apa-get-ActiveSelectionSet Object)
```

Arguments

Object	The Drawing object to which this property applies.

Returns

The active selection set for the drawing.

ActiveSpace Property

Controls the active space between paper space and model space.

```
(apa-get-ActiveSpace Object)
(apa-set-ActiveSpace Object ActiveSpace)
```

Arguments

Object	The Drawing object to which this property applies.
ActiveSpace	**acModelSpace** = 1
	acPaperSpace = 0

Returns

An integer specifying the active space for the drawing.

Remarks

This property is stored in the TILEMODE system variable.

ActiveTextStyle Property

Specifies the active text style for the drawing.

```
(apa-get-ActiveTextStyle Object)
(apa-set-ActiveTextStyle Object ActiveTextStyle)
```

Arguments

Object   The Drawing object to which this property applies.

ActiveTextStyle   The TextStyle object to become the active paper space viewport.

Returns

The active text style as a TextStyle object.

ActiveUCS Property

Specifies the active UCS for the drawing.

```
(apa-get-ActiveUCS Object)
(apa-set-ActiveUCS Object ActiveUCS)
```

Arguments

Object   The Drawing object to which this property applies.

ActiveUCS   The UCS object to become the active UCS.

Returns

The UCS as a UCS object.

ActiveViewport Property

Specifies the active viewport for the drawing.

```
(apa-get-ActiveViewport Object)
(apa-set-ActiveViewport Object ActiveViewport)
```

Arguments

Object   The Drawing object to which this property applies.

ActiveViewport The Viewport object to become the active Viewport.

Returns

The Viewport as a Viewport object.

ADCInsertUnitsDefaultSource

Specifies the Autodesk Design Center default insertion units for a source drawing.

```
(apa-get-ADCInsertUnitsDefaultSource Object)
(apa-set-ADCInsertUnitsDefaultSource Object InsertUnits)
```

Arguments

Object   The PreferencesUser object to which this property applies.

InsertUnits An integer specifying the default insertion units. One of the following:

`acInsertUnitsUnitless`
`acInsertUnitsInches`
`acInsertUnitsFeet`
`acInsertUnitsMiles`
`acInsertUnitsMillimeters`
`acInsertUnitsCentimeters`
`acInsertUnitsMeters`
`acInsertUnitsKilometers`

```
acInsertUnitsMicroinches
acInsertUnitsMils
acInsertUnitsYards
acInsertUnitsAngstroms
acInsertUnitsNanometers
acInsertUnitsMicrons
acInsertUnitsDecimeters
acInsertUnitsDecameters
acInsertUnitsHectometers
acInsertUnitsGigameters
acInsertUnitsAstronomicalUnits
acInsertUnitsLightYears
acInsertUnitsParsecs
```

Returns

An integer specifying the Autodesk Design Center default insertion units for a source drawing.

Remarks

This property is stored in the INSUNITSDEFSOURCE system variable.

ADCInsertUnitsDefaultTarget

Specifies the Autodesk Design Center default insertion units for a target drawing.

```
(apa-get-ADCInsertUnitsDefaultTarget Object)
(apa-set-ADCInsertUnitsDefaultTarget Object InsertUnits)
```

Arguments

Object	The PreferencesUser object to which this property applies.
InsertUnits	An integer specifying the default insertion units. One of the following:

```
acInsertUnitsUnitless
acInsertUnitsInches
acInsertUnitsFeet
acInsertUnitsMiles
acInsertUnitsMillimeters
acInsertUnitsCentimeters
acInsertUnitsMeters
acInsertUnitsKilometers
acInsertUnitsMicroinches
acInsertUnitsMils
acInsertUnitsYards
acInsertUnitsAngstroms
acInsertUnitsNanometers
acInsertUnitsMicrons
acInsertUnitsDecimeters
acInsertUnitsDecameters
acInsertUnitsHectometers
acInsertUnitsGigameters
acInsertUnitsAstronomicalUnits
acInsertUnitsLightYears
acInsertUnitsParsecs
```

Returns

An integer specifying the Autodesk Design Center default insertion units for a target drawing.

Remarks

This property is stored in the INSUNITSDEFTARGET system variable.

Alignment Property

Specifies the vertical and horizontal alignments for the attribute, attribute reference, or text.

```
(apa-get-Alignment Object)
(apa-set-Alignment Object Alignment)
```

Arguments

Object　　　　　　The Attribute, AttributeRef, or Text object to which this property applies.

Alignment　　　　An integer specifying the alignment. One of the following:

```
acAlignmentLeft
acAlignmentCenter
acAlignmentRight
acAlignmentAligned
acAlignmentMiddle
acAlignmentFit
acAlignmentTopLeft
acAlignmentTopCenter
acAlignmentTopRight
acAlignmentMiddleLeft
acAlignmentMiddleCenter
acAlignmentMiddleRight
acAlignmentBottomLeft
acAlignmentBottomCenter
acAlignmentBottomRight
```

Returns

The alignment as an integer.

Remarks

Text aligned to **acAlignmentLeft** uses only the InsertionPoint property to locate the text.

You cannot set the TextAlignmentPoint property while the text is aligned to **acAlignmentLeft**.

Text aligned to **acAlignmentAligned**, or **acAlignmentFit** uses both the InsertPoint and TextAlignmentPoint properties.

All other alignments use only the TextAlignmentPoint property to locate the text.

AlignmentPointAcquisition Property

Specifies how AutoAlignment points are acquired.

```
(apa-get-AlignmentPointAcquisition Object)
(apa-set-AlignmentPointAcquisition Object Mode)
```

Arguments

Object　　　　　　The PreferencesDrafting object to which this property applies.

Mode　　　　　　　An integer specifying the alignment point acquisition mode. One of the following:

```
acAlignPntAcquisitionAutomatic
acAlignPntAcquisitionShiftToAcquire
```

Returns

The mode as an integer.

AllowLongSymbolNames Property

Determines if symbol names may include extended character sets and more than 31 characters.

```
(apa-get-AllowLongSymbolNames Object)
(apa-set-AllowLongSymbolNames Object Allow)
```

Arguments

Object	The DatabasePreferences object to which this property applies.
Allow	non-nil to allow long symbol names; nil to disallow.

Returns

T if long symbol names are allowed, nil if they're not.

Remarks

This property is stored in the EXTNAMES system variable.

Long symbol names may contain up to 255 characters not used by Microsoft Windows or AutoCAD for other purposes. This is the default for AutoCAD 2000. Long symbol names were not allowed in AutoCAD R14 and earlier.

Short symbol names may contain up to 31 characters [A-Z, 0-9, $, _, and hyphen -.

This property applies to the names of non-graphical objects stored in symbol tables.

AltFontFile Property

Specifies the font file to use if AutoCAD cannot locate the original font or an alternative from the font-mapping file.

```
(apa-get-AltFontFile Object)
(apa-set-AltFontFile Object FileName)
```

Arguments

Object	The PreferencesFiles object to which this property applies.
FileName	A string, specifying the name of the alternate font file.

Returns

A string, specifying the name of the alternate font file.

AltRoundDistance Property

Specifies the rounding of alternate units for dimensions.

```
(apa-get-AltRoundDistance Object)
(apa-set-AltRoundDistance Object RoundDistance)
```

Arguments

Object	The DimAligned, DimDiametric, DimOrdinate, DimRadial, or DimRotated object to which this property applies.
RoundDistance	A positive number specifying the value to which to round distances, when the AltUnits property is on.

Returns

A real number.

Remarks

Overrides the DIMALTRND system variable for the specified dimension.

AltSuppressLeadingZeros Property

Specifies the suppression of leading zeros in alternate dimension values.

```
(apa-get-AltSuppressLeadingZeros Object)
(apa-set-AltSuppressLeadingZeros Object Suppress)
```

Arguments

Object	The DimAligned, DimDiametric, DimOrdinate, DimRadial, or DimRotated object to which this property applies.
Suppress	non-nil to suppress leading zeros. nil to not-suppress.

Returns

T if leading zeros are suppressed. nil otherwise.

Remarks

Overrides the DIMALTZ system variable for the specified dimension.

AltSuppressTrailingZeros Property

Specifies the suppression of trailing zeros in alternate dimension values.

```
(apa-get-AltSuppressTrailingZeros Object)
(apa-set-AltSuppressTrailingZeros Object Suppress)
```

Arguments

Object   The DimAligned, DimDiametric, DimOrdinate, DimRadial, or DimRotated object to
                which this property applies.

Suppress   non-nil to suppress trailing zeros. nil to not-suppress.

Returns

T if trailing zeros are suppressed, nil otherwise.

Remarks

Overrides the DIMALTZ system variable for the specified dimension.

AltSuppressZeroFeet Property

Specifies the suppression of a zero foot measurement in alternate dimension values.

```
(apa-get-AltSuppressZeroFeet Object)
(apa-set-AltSuppressZeroFeet Object Suppress)
```

Arguments

Object   The DimAligned, DimDiametric, DimOrdinate, DimRadial, or DimRotated object to
                which this property applies.

Suppress   non-nil to suppress zero-feet. nil to not-suppress.

Returns

T if zero feet are suppressed. nil otherwise.

Remarks

Overrides the DIMALTZ system variable for the specified dimension.

AltSuppressZeroInches Property

Specifies the suppression of a zero inch measurement in alternate dimension values.

```
(apa-get-AltSuppressZeroInches Object)
(apa-set-AltSuppressZeroInches Object Suppress)
```

Arguments

Object   The DimAligned, DimDiametric, DimOrdinate, DimRadial, or DimRotated object to
                which this property applies.

Suppress   non-nil to suppress zero inches, nil to not-suppress.

Returns

T if zero inches are suppressed. nil otherwise.

Remarks

Overrides the DIMALTZ system variable for the specified dimension.

AltTabletMenuFile Property

Specifies the path for an alternate menu to swap with the standard AutoCAD tablet menu.

```
(apa-get-AltTabletMenuFile Object)
(apa-set-AltTabletMenuFile Object FileName)
```

Arguments

Object	The PreferencesFiles object to which this property applies.
FileName	A string, specifying the name of the tablet menu file.

Returns

A string, specifying the name of the alternate tablet Menu file.

AltTextPrefix Property

Specifies a prefix for the alternate dimension measurement.

```
(apa-get-AltTextPrefix Object)
(apa-set-AltTextPrefix Object Prefix)
```

Arguments

Object	The DimAligned, DimDiametric, DimOrdinate, DimRadial, or DimRotated object to which this property applies.
Prefix	A string, specifying the prefix for the alternate dimensions.

Returns

A string, specifying the prefix for the alternate dimensions.

Remarks

Overrides the DIMAPOST system variable for the specified dimension.

AltTextSuffix Property

Specifies a suffix for the alternate dimension measurement.

```
(apa-get-AltTextSuffix Object)
(apa-set-AltTextSuffix Object Suffix)
```

Arguments

Object	The DimAligned, DimDiametric, DimOrdinate, DimRadial, or DimRotated object to which this property applies.
Suffix	A string, specifying suffix for the alternate dimensions.

Returns

A string, specifying suffix for the alternate dimensions.

Remarks

Overrides the DIMAPOST system variable for the specified dimension.

AltTolerancePrecision Property

Specifies the precision of tolerance values in alternate dimensions.

```
(apa-get-AltTolerancePrecision Object)
(apa-set-AltTolerancePrecision Object Precision)
```

Arguments

Object	The DimAligned, DimDiametric, DimOrdinate, DimRadial, or DimRotated object to which this property applies.

Precision	An integer specifying the precision. One of the following:	
	acDimPrecisionZero:	0
	acDimPrecisionOne:	0.0
	acDimPrecisionTwo:	0.00
	acDimPrecisionThree:	0.000
	acDimPrecisionFour:	0.0000
	acDimPrecisionFive:	0.00000
	acDimPrecisionSix:	0.000000
	acDimPrecisionSeven:	0.0000000
	acDimPrecisionEight:	0.00000000

Returns

An integer.

Remarks

Overrides the DIMALTTD system variable for the specified dimension.

AltToleranceSuppressLeadingZeros Property

Specifies the suppression of leading zeros in alternate tolerance dimension values.

```
(apa-get-AltToleranceSuppressLeadingZeros Object)
(apa-set-AltToleranceSuppressLeadingZeros Object Suppress)
```

Arguments

Object	The DimAligned, DimDiametric, DimOrdinate, DimRadial, or DimRotated object to which this property applies.
Suppress	non-nil to suppress leading zeros. nil to not-suppress

Returns

T if leading zeros are suppressed. nil otherwise.

Remarks

Overrides the DIMALTTZ system variable for the specified dimension.

AltToleranceSuppressTrailingZeros Property

Specifies the suppression of trailing zeros in alternate tolerance dimension values.

```
(apa-get-AltToleranceSuppressTrailingZeros Object)
(apa-set-AltToleranceSuppressTrailingZeros Object Suppress)
```

Arguments

Object	The DimAligned, DimDiametric, DimOrdinate, DimRadial, or DimRotated object to which this property applies.
Suppress	non-nil to suppress trailing zeros. nil to not-suppress

Returns

T if trailing zeros are suppressed. nil otherwise.

Remarks

Overrides the DIMALTTZ system variable for the specified dimension.

AltToleranceSuppressZeroFeet Property

Specifies the suppression of a zero foot measurement in alternate tolerance values.

```
(apa-get-AltToleranceSuppressZeroFeet Object)
(apa-set-AltToleranceSuppressZeroFeet Object Suppress)
```

Arguments

Object	The DimAligned, DimDiametric, DimOrdinate, DimRadial, or DimRotated object to which this property applies.
Suppress	non-`nil` to suppress zero-feet. `nil` to not-suppress.

Returns

`T` if zero feet are suppressed. `nil` otherwise.

Remarks

Overrides the DIMALTTZ system variable for the specified dimension.

AltToleranceSuppressZeroInches Property

Specifies the suppression of a zero inch measurement in alternate tolerance values.

```
(apa-get-AltToleranceSuppressZeroInches Object)
(apa-set-AltToleranceSuppressZeroInches Object Suppress)
```

Arguments

Object	The DimAligned, DimDiametric, DimOrdinate, DimRadial, or DimRotated object to which this property applies.
Suppress	non-`nil` to suppress zero inches, `nil` to not-suppress.

Returns

`T` if zero inches are suppressed. `nil` otherwise.

Remarks

Overrides the DIMALTTZ system variable for the specified dimension.

AltUnits Property

Enables alternate units dimensioning.

```
(apa-get-AltUnits Object)
(apa-set-AltUnits Object Enable)
```

Arguments

Object	The DimAligned, DimDiametric, DimOrdinate, DimRadial, or DimRotated object to which this property applies.
Enable	non-`nil` to enable alternate units, `nil` to not-suppress.

Returns

`T` if alternate units are enabled. `nil` otherwise.

Remarks

Overrides the DIMALT system variable for the specified dimension.

AltUnitsFormat Property

Specifies the format of values in alternate dimensions.

```
(apa-get-AltUnitsFormat Object)
(apa-set-AltUnitsFormat Object Format)
```

Arguments

Object	The DimAligned, DimDiametric, DimOrdinate, DimRadial, or DimRotated object to which this property applies.

Format	An integer specifying the format. One of the following:
	`acDimScientific`
	`acDimDecimal`
	`acDimEngineering`
	`acDimArchitecturalStacked`
	`acDimFractionalStacked`
	`acDimArchitectural`
	`acDimFractional`
	`acDimWindowsDesktop`

Returns

An integer.

Remarks

Overrides the DIMALTU system variable for the specified dimension.

AltUnitsPrecision Property

Specifies the precision of alternate units values in dimensions.

```
(apa-get-AltUnitsPrecision Object)
(apa-set-AltUnitsPrecision Object Precision)
```

Arguments

Object	The DimAligned, DimDiametric, DimOrdinate, DimRadial, or DimRotated object to which this property applies.
Precision	An integer specifying the precision. One of the following:

`acDimPrecisionZero:`	0
`acDimPrecisionOne:`	0.0
`acDimPrecisionTwo:`	0.00
`acDimPrecisionThree:`	0.000
`acDimPrecisionFour:`	0.0000
`acDimPrecisionFive:`	0.00000
`acDimPrecisionSix:`	0.000000
`acDimPrecisionSeven:`	0.0000000
`acDimPrecisionEight:`	0.00000000

Returns

An integer.

Remarks

Overrides the DIMALTD system variable for the specified dimension.

AltUnitsScale Property

Specifies the scale factor for alternate units dimensions.

```
(apa-get-AltUnitsScale Object)
(apa-set-AltUnitsScale Object AltUnitsScale)
```

Arguments

Object	The DimAligned, DimDiametric, DimOrdinate, DimRadial, or DimRotated object to which this property applies.
AltUnitsScale	A positive number specifying the scale factor.

Returns

A real number.

Remarks

Overrides the DIMALTF system variable for the specified dimension.

Angle Property

Returns the angle of a line.

`(apa-get-Angle ` *`Object`*`)`

Arguments

Object   The Line object to be queried.

Returns

A real number representing the angle of the line from the X-axis, in radians.

AngleFormat Property

Specifies the unit format for angular dimensions.

`(apa-get-AngleFormat ` *`Object`*`)`
`(apa-set-AngleFormat ` *`Object AngleFormat`*`)`

Arguments

Object   The Dim3PointAngular or DimAngular object to which this property applies.

AngleFormat   An integer specifying the angle format. One of the following:

acDegrees
acDegreeMinuteSeconds
acGrads
acRadians
acDim

Returns

An integer.

Remarks

This property overrides the value of the DIMAUNIT system variable for the given dimension.

AngleVertex Property

Specifies the angle vertex for a three point angular dimension.

`(apa-get-AngleVertex ` *`Object`*`)`
`(apa-set-AngleVertex ` *`Object AngleVertex`*`)`

Arguments

Object   The Dim3PointAngular object to which this property applies.

AngleVertex   A 2D or 3D point specifying the angle vertex.

Returns

A 3D point.

Annotation Property

Specifies the annotation object for a leader.

`(apa-get-Annotation ` *`Object`*`)`
`(apa-set-Annotation ` *`Object Annotation`*`)`

Arguments

Object   The Leader object to which this property applies.

Annotation   The Tolerance, MText, or BlockRef annotation object for the leader.

Returns

The Annotation object attached to the leader.

Application Property

Returns the Application object.

```
(apa-get-Application Object)
```

Arguments

Object   Any object.

Returns

The Application object.

ArcLength Property

Returns the length of an arc.

```
(apa-get-ArcLength Object)
```

Arguments

Object   The Arc object to be queried.

Returns

A real number.

ArcSmoothness Property

Specifies the smoothness percentage of circles, arcs, and ellipses.

```
(apa-get-ArcSmoothness Object)
(apa-set-ArcSmoothness Object Smoothness)
```

Arguments

Object   The PViewport or Viewport object to which this property applies.

Smoothness   An integer in the range 1 to 20,000. The default is 100.

Returns

An integer.

Area Property

Specifies the enclosed area of an object.

```
(apa-get-Area Object)
(apa-set-Area Object Area)
```

Arguments

Object   The Arc, Circle, Ellipse, LightweightPolyline, Polyline, Region, or Spline object to which this property applies.

Area   A positive number, specifying the area of the object, in square drawing units.

Returns

A real number, or nil if not successful.

Arrowhead1Block Property

Specifies the block to use as the custom arrowhead for the first end of the dimension line.

```
(apa-get-Arrowhead1Block Object)
(apa-set-Arrowhead1Block Object Arrowhead1Block)
```

Arguments

object   Tee Dim3PointAngular, DimAligned, DimAngular, DimDiametric, or DimRotated object to which this property applies.

Arrowhead1Block   A string, specifying the name of the block to use as the arrowhead for the first end of the dimension line. Use "." to specify no block is to be used.

Returns

A string.

Remarks

This property overrides the value of the DIMBLK1 system variable for the given dimension.

Arrowhead1Type Property

Specifies the type of arrowhead for the first end of the dimension line.

```
(apa-get-Arrowhead1Type Object)
(apa-set-Arrowhead1Type Object ArrowType)
```

Arguments

Object	The Dim3PointAngular, DimAligned, DimAngular, DimDiametric, or DimRotated object to which this property applies.
ArrowType	An integer specifying the type of arrow head. One of the following:

```
acArrowDefault
acArrowDot
acArrowDotSmall
acArrowDotBlank
acArrowOrigin
acArrowOrigin2
acArrowOpen
acArrowOpen90
acArrowOpen30
acArrowClosed
acArrowSmall
acArrowNone
acArrowOblique
acArrowBoxFilled
acArrowBoxBlank
acArrowClosedBlank
acArrowDatumFilled
acArrowDatumBlank
acArrowIntegral
acArrowArchTick
acArrowUserDefined
```

Returns

An integer.

Remarks

This property overrides the value of the DIMBLK1 system variable for the given dimension.

Arrowhead2Block Property

Specifies the block to use as the custom arrowhead for the first end of the dimension line.

```
(apa-get-Arrowhead2Block Object)
(apa-set-Arrowhead2Block Object Arrowhead2Block)
```

Arguments

object	The Dim3PointAngular, DimAligned, DimAngular, DimDiametric, or DimRotated object to which this property applies.
Arrowhead2Block	A string, specifying the name of the block to use as the arrowhead for the first end of the dimension line. Use "." to specify no block is to be used.

Returns

A string.

Remarks

This property overrides the value of the DIMBLK2 system variable for the given dimension.

Arrowhead2Type Property

Specifies the type of arrowhead for the first end of the dimension line.

```
(apa-get-Arrowhead2Type Object)
(apa-set-Arrowhead2Type Object ArrowType)
```

Arguments

Object The Dim3PointAngular, DimAligned, DimAngular, DimDiametric, or DimRotated object to which this property applies.

ArrowType An integer specifying the type of arrow head. One of the following:

```
acArrowDefault
acArrowDot
acArrowDotSmall
acArrowDotBlank
acArrowOrigin
acArrowOrigin2
acArrowOpen
acArrowOpen90
acArrowOpen30
acArrowClosed
acArrowSmall
acArrowNone
acArrowOblique
acArrowBoxFilled
acArrowBoxBlank
acArrowClosedBlank
acArrowDatumFilled
acArrowDatumBlank
acArrowIntegral
acArrowArchTick
acArrowUserDefined
```

Returns

An integer.

Remarks

This property overrides the value of the DIMBLK2 system variable for the given dimension.

ArrowheadBlock Property

Specifies the block to use as the custom arrowhead for a radial dimension or leader line.

```
(apa-get-Arrowhead1Block Object)
(apa-set-Arrowhead1Block Object ArrowheadBlock)
```

Arguments

Object The DimRadial or Leader object to which this property applies.

ArrowheadBlock A string, specifying the name of the block to use as the arrowhead for the first end of the dimension line. Use "." to specify no block is to be used.

Returns

A string.

Remarks

This property overrides the value of the DIMLDRBLK system variable for the given dimension.

ArrowheadSize Property

Specifies the size of dimension line arrowheads and leader line arrowheads.

```
(apa-get-ArrowheadSize Object)
(apa-set-ArrowheadSize Object Size)
```

Arguments

Object	The Dim3PointAngular, DimAligned, DimAngular, DimDiametric, DimOrdinate, DimRadial, DimRotated, or Leader object to which this property applies.
Size	A positive number specifying the arrowhead size.

Returns

A real number.

Remarks

This property overrides the value of the DIMASZ system variable for the given dimension.

ArrowheadType Property

Specifies the type of arrowhead for a radial dimension or leader.

```
(apa-get-ArrowheadType Object)
(apa-set-ArrowheadType Object ArrowType)
```

Arguments

Object	The Dim3PointAngular, DimAligned, DimAngular, DimDiametric, or DimRotated object to which this property applies.
ArrowType	An integer specifying the type of arrow head. One of the following:

```
acArrowDefault
acArrowDot
acArrowDotSmall
acArrowDotBlank
acArrowOrigin
acArrowOrigin2
acArrowOpen
acArrowOpen90
acArrowOpen30
acArrowClosed
acArrowSmall
acArrowNone
acArrowOblique
acArrowBoxFilled
acArrowBoxBlank
acArrowClosedBlank
acArrowDatumFilled
acArrowDatumBlank
acArrowIntegral
acArrowArchTick
acArrowUserDefined
```

Returns

An integer.

Remarks

This property overrides the value of the DIMLDRBLK system variable for the given dimension.

AssociativeHatch Property

Determines if the hatch is associative or not.

```
(apa-get-AssociativeHatch Object)
```

Arguments

Object   The Hatch object to be queried.

Returns

T if the hatch is associative, nil if it's not.

Remarks

Associativity can be set when only as a hatch is created.

AttachmentPoint Property

Specifies the attachment point (alignment) for an MText object.

```
(apa-get-AttachmentPoint Object)
(apa-set-AttachmentPoint Object AttachmentPoint)
```

Arguments

Object   The MText object to which this property applies.

AttachmentPoint   An integer specifying the alignment of the MText within the text boundary. One of the following:

```
acAttachmentPointTopLeft
acAttachmentPointTopCenter
acAttachmentPointTopRight
acAttachmentPointMiddleLeft
acAttachmentPointMiddleCenter
acAttachmentPointMiddleRight
acAttachmentPointBottomLeft
acAttachmentPointBottomCenter
acAttachmentPointBottomRight
```

Returns

An integer.

Remarks

When you change the AttachmentPoint property, the position of the bounding box does not change; the text is re-justified within the bounding box.. The value of the InsertionPoint property will change accordingly.

AutoAudit Property

Specifies if AutoCAD should perform an audit after you render a DXFIN or DXBIN interchange command.

```
(apa-get-AutoAudit Object)
(apa-set-AutoAudit Object AutoAudit)
```

Arguments

Object   The PreferencesOpenSave object to which this property applies.

AutoAudit   non-nil performs an audit after each DXFIN or DXBIN command. nil does not.

Returns

T if AutoAudit is enabled. nil otherwise.

AutoSaveInterval Property

Specifies the automatic save interval in minutes.

```
(apa-get-AutoSaveInterval Object)
(apa-set-AutoSaveInterval Object AutoSaveInterval)
```

Arguments

Object	The PreferencesOpenSave object to which this property applies.
AutoSaveInterval	An integer in the range $0 \leq AutoSaveInterval \leq 600$.

Returns

An Integer.

Remarks

This value of this property is stored in the SAVETIME system variable.

AutoSavePath Property

Specifies the path for automatic save files.

```
(apa-get-AutoSavePath Object)
(apa-set-AutoSavePath Object AutoSavePath)
```

Arguments

Object	The PreferencesFiles object to which this property applies.
AutoSavePath	A string, specifying the path for the automatic save files.

Returns

A string.

Remarks

This value of this property is stored in the SAVEFILEPATH system variable.

AutoSnapAperture Property

Controls the display of the AutoSnap aperture.

```
(apa-get-AutoSnapAperture Object)
(apa-set-AutoSnapAperture Object Enable)
```

Arguments

Object	The PreferencesDrafting object to which this property applies.
Enable	non-nil enables, nil disables.

Returns

T if enabled. nil otherwise.

Remarks

This value of this property is stored in the APBOX system variable.

AutoSnapApertureSize Property

Specifies the size of the AutoSnap aperture.

```
(apa-get-AutoSnapApertureSize Object)
(apa-set-AutoSnapApertureSize Object Size)
```

Arguments

Object	The PreferencesDrafting object to which this property applies.
Size	An integer specifying the size in pixels.

Returns

An Integer.

Remarks

This value of this property is stored in the APERTURE system variable.

AutoSnapMagnet Property

Controls the AutoSnap magnet.

```
(apa-get-AutoSnapMagnet Object)
(apa-set-AutoSnapMagnet Object Enable)
```

Arguments

Object	The PreferencesDrafting object to which this property applies.
Enable	non-nil enables, nil disables.

Returns

T if enabled. nil otherwise.

Remarks

This value of this property is stored in the AUTOSNAP system variable.

AutoSnapMarker Property

Controls the AutoSnap marker.

```
(apa-get-AutoSnapMarker Object)
(apa-set-AutoSnapMarker Object Enable)
```

Arguments

Object	The PreferencesDrafting object to which this property applies.
Enable	non-nil enables, nil disables.

Returns

T if enabled. nil otherwise.

Remarks

This value of this property is stored in the AUTOSNAP system variable.

AutoSnapMarkerColor Property

Specifies the color of the AutoSnap marker.

```
(apa-get-AutoSnapMarkerColor Object)
(apa-set-AutoSnapMarkerColor Object Color)
```

Arguments

Object	The PreferencesDrafting object to which this property applies.
Color	An integer specifying the ACI of the AutoSnap marker. One of the following:

> acRed
> acYellow
> acGreen
> acCyan
> acBlue
> acMagenta
> acWhite

Returns

An Integer.

AutoSnapMarkerSize Property

Specifies the size of the AutoSnap marker.

```
(apa-get-AutoSnapMarkerSize Object)
(apa-set-AutoSnapMarkerSize Object Size)
```

Arguments

Object   The PreferencesDrafting object to which this property applies.

Size   An integer in the range $1 \leq Size \leq 20$, specifying the size in pixels.

Returns

An Integer.

AutoSnapToolTip Property

Controls display of the AutoSnap ToolTips.

```
(apa-get-AutoSnapToolTip Object)
(apa-set-AutoSnapToolTip Object Enable)
```

Arguments

Object   The PreferencesDrafting object to which this property applies.

Enable   non-nil enables, nil disables.

Returns

T if enabled. nil otherwise.

Remarks

This value of this property is stored in the AUTOSNAP system variable.

AutoTrackToolTip Property

Controls the display of the AutoTrack ToolTips.

```
(apa-get-AutoTrackToolTip Object)
(apa-set-AutoTrackToolTip Object Enable)
```

Arguments

Object   The PreferencesDrafting object to which this property applies.

Enable   non-nil enables, nil disables.

Returns

T if enabled. nil otherwise.

Remarks

This value of this property is stored in the AUTOSNAP system variable.

AutoTrackingVecColor Property

Specifies the color of the AutoTracking vector.

```
(apa-get-AutoTrackingVecColor Object)
(apa-set-AutoTrackingVecColor Object Color)
```

Arguments

Object   The PreferencesDisplay object to which this method applies.

Color   An integer specifying an OLE_COLOR. One of the following:

    vbBlack
    vbRed
    vbYellow
    vbGreen
    vbCyan
    vbBlue
    vbMagenta
    vbWhite

Returns

An integer.

Backward Property

Specifies the direction of text.

```
(apa-get-Backward Object)
(apa-set-Backward Object Mode)
```

Arguments

Object	The Attribute, AttributeReference, or Text object to which this property applies.
Mode	non-nil for backward, nil for forward.

Returns

T if backward. nil for forward.

BasePoint Property

Specifies the point through which the ray or xline passes.

```
(apa-get-BasePoint Object)
(apa-set-BasePoint Object BasePoint)
```

Arguments

Object	The Ray or XLine object to which this property applies.
BasePoint	The point through which the ray or xline passes.

Returns

A 3D point.

BatchPlotProgress Property

Returns the status of the batch plot, or terminates the batch plot.

```
(apa-get-BatchPlotProgress Object)
(apa-set-BatchPlotProgress Object Progress)
```

Arguments

Object	The Plot object to which this property applies.
Progress	Setting *Progress* to nil will terminate the batch plot.

Returns

T if the batch plot is in progress, nil otherwise.

BeepOnError Property

Specifies if AutoCAD should beep when it detects an invalid entry.

```
(apa-get-BeepOnError Object)
(apa-set-BeepOnError Object Enable)
```

Arguments

Object	The PreferencesSystem object to which this property applies.
Enable	non-nil enables, nil disables.

Returns

T if enabled. nil otherwise.

BigFontFile Property

Specifies the name of the big font file associated with the text or attribute.

```
(apa-get-BigFontFile Object)
(apa-set-BigFontFile Object FileName)
```

Arguments

`Object`	The TextStyle object to which this property applies.
`FileName`	String, specifying the name and extension the big font file.

Returns

A string.

Remarks

The only valid file type is *.shx.*

This property cannot be set to `nil` or an empty string.

Block Property

Returns the block associated with the layout.

```
(apa-get-Block Object)
```

Arguments

`Object`	The Layout object to which this property applies.

Returns

The block object associated with the layout.

Blocks Property

Returns the Blocks collection for the specified drawing.

```
(apa-get-Blocks Object)
```

Arguments

`Object`	The Document object to which this property applies.

Returns

The blocks collection object associated with the drawing.

Brightness Property

Specifies the current brightness value of an image.

```
(apa-get-Brightness Object)
(apa-set-Brightness Object Brightness)
```

Arguments

`Object`	The Raster object to which this property applies.
`Brightness`	An integer in the range $0 \leq Brightness \leq 100$ specifying the brightness of the image.

Returns

An integer.

Remarks

Bi-tonal images cannot be adjusted.

You can adjust the brightness, contrast, and fade properties of an image to control its display as well as its plotted output. Doing so does not affect the original raster image file.

The brightness property darkens or lightens an image.

The contrast property can make images with poor quality easier to read.

The fade property can makes vectors easier to see over images, and can create a watermark effect in the plotted output.

CanonicalMediaName Property

Specifies the paper size by name.

```
(apa-get-CanonicalMediaName Object)
(apa-set-CanonicalMediaName Object Name)
```

Arguments

Object   The Layout or PlotConfiguration object to which this method applies.

Name   A string, specifying the name of the paper size.

Returns

A string.

Remarks

Changes will not be visible until the drawing is regenerated.

Caption Property

Returns the text that the user sees displayed for the application or menu item.

```
(apa-get-Caption Object)
```

Arguments

Object   The Application or PopupMenuItem object to which this method applies.

Returns

A string

Remarks

For a menu item, this property is read-only and is derived from the Label property by removing any DIESEL string expressions.

Center Property

Specifies the center of an arc, circle, ellipse, view, or viewport.

```
(apa-get-Center Object)
(apa-set-Center Object Center)
```

Arguments

Object   The Arc, Circle, Ellipse, PViewport, View, or Viewport object to which this method applies.

Center   Point specifying the center of the object.

Returns

A point.

CenterMarkSize Property

Specifies the size of the center mark for radial and diameter dimensions.

```
(apa-get-CenterMarkSize Object)
(apa-set-CenterMarkSize Object CenterMarkSize)
```

Arguments

Object   The DimDiametric or DimRadial object to which this method applies.

Size   A positive number specifying the size of the center mark.

Returns

A real number.

Remarks

This property overrides the value of the DIMCEN system variable for the given dimension.

This property is not available if the CenterType property is set to **acCenterNone**.

CenterPlot Property

Specifies the centering of the plot on the media.

```
(apa-get-CenterPlot Object)
(apa-set-CenterPlot Object CenterPlot)
```

Arguments

Object   The Layout or PlotConfiguration object to which this property applies.

CenterPlot   non-nil centers the plot, nil doesn't.

Returns

T if centered. nil otherwise.

Remarks

Changes will not be visible until the drawing is regenerated.

This property cannot be set to T on a layout object whose PlotType property is set to acLayout.

CenterType Property

Specifies the type of the center mark for radial and diameter dimensions.

```
(apa-get-CenterType Object)
(apa-set-CenterType Object CenterType)
```

Arguments

Object   The DimDiametric or DimRadial object to which this method applies.

CenterType   A integer specifying the type of center mark. One of the following:

             acCenterMark
             acCenterLine
             acCenterNone

Returns

An integer.

Remarks

This property overrides the value of the DIMCEN system variable for the given dimension.

Centroid Property

Returns the center of area or mass for a region or solid.

```
(apa-get-Centroid Object)
```

Arguments

Object   The Region or 3DSolid object to which this property applies.

Returns

A 3D point for a 3DSolid, a 2D point for a region

Check Property

Specifies the check status for the popup menu item.

```
(apa-get-Check Object)
(apa-set-Check Object Check)
```

Arguments

object	The PopupMenuItem object to which this property applies.
Check	non-nil to check the menu item, nil to un-check it.

Returns

T if the menu item is checked, nil it isn't.

Circumference Property

Specifies the circumference of a circle.

```
(apa-get-Circumference Object)
(apa-set-Circumference Object Circumference)
```

Arguments

Object	The Circle object to which this property applies.
Circumference	A positive number, specifying the circumference.

Returns

A real number.

Clipped Property

Determines if the viewport has been clipped.

```
(apa-get-Clipped Object)
```

Arguments

Object	The PViewport object to which this property applies.

Returns

T if the viewport is clipped, nil otherwise.

ClippingEnabled Property

Specifies if clipping is enabled for the raster image.

```
(apa-get-ClippingEnabled Object)
(apa-set-ClippingEnabled Object Enable)
```

Arguments

Object	The Raster object to which this property applies.
Enable	non-nil to enable clipping, nil to disable it

Returns

T if clipping is enabled, nil if it's not.

Closed Property

Specifies if the 3D polyline, lightweight polyline, polyline, or spline is open or closed.

```
(apa-get-Closed Object)
(apa-set-Closed Object Closed)
```

Arguments

Object	The 3DPoly, LightweightPolyline, Polyline, or Spline object to which this property applies.
Closed	non-nil to close the object, nil to open it

Returns

T if the object is closed, nil otherwise.

Color Property

Specifies the color of an entity or layer.

```
(apa-get-Color Object)
(apa-set-Color Object Color)
```

Arguments

Object	The All Drawing Objects, Group, or Layer object to which this property applies.
Color	An ACI integer in the range $0 \leq Color \leq 256$, or one of the following:

> **acByBlock**
> **acByLayer**
> **acRed**
> **acYellow**
> **acGreen**
> **acCyan**
> **acBlue**
> **acMagenta**
> **acWhite**

Returns

An integer.

Columns Property

Specifies the number of columns in a block array.

```
(apa-get-Columns Object)
(apa-set-Columns Object ColumnSpacing)
```

Arguments

Object	The MInsertBlock object to which this property applies.
Columns	Positive number specifying the number of columns in the block array.

Returns

An integer.

ColumnSpacing Property

Specifies the spacing of columns in a block array.

```
(apa-get-Columns Object)
(apa-set-Columns Object Spacing)
```

Arguments

Object	The MInsertBlock object to which this property applies.
Spacing	Number specifying the spacing of columns in the block array.

Returns

An real number.

ConfigFile Property

Returns the location of the configuration file used to store hardware device driver information.

```
(apa-get-ConfigFile Object)
```

Arguments

Object	The PreferencesFiles object to which this property applies.

Returns

A string.

ConfigName Property

Specifies the plotter configuration name.

```
(apa-get-ConfigName Object)
(apa-set-ConfigName Object ConfigName)
```

Arguments

Object   The Layout or PlotConfiguration object to which this property applies.

ConfigName   The name of the *.pc3* file or print device to be used by the layout or plot configuration.

Returns

A string.

Constant Property

Specifies whether or not the attribute or attribute reference is constant.

```
(apa-get-Constant Object)
(apa-set-Constant Object Constant)
```

Arguments

Object   The Attribute or AttributeReference object to which this property applies.

Constant   non-nil to set the attribute as constant. nil to set it as non-constant.

Returns

T if the object is constant, nil if it is not.

ConstantWidth Property

Specifies a global width for all segments in a polyline.

```
(apa-get-ConstantWidth Object)
(apa-set-ConstantWidth Object Width)
```

Arguments

Object   The LightweightPolyline or Polyline object to which this property applies.

Width   A number specifying the constant width.

Returns

A real number, or nil if the polyline does not have a ConstantWidth property.

ContourlinesPerSurface Property

Specifies the number of contour lines (isolines) per surface on 3D Solid objects.

```
(apa-get-ContourlinesPerSurface Object)
(apa-set-ContourlinesPerSurface Object Isolines)
```

Arguments

Object   The DatabasePreferences object to which this property applies.

Isolines   An integer in the range $0 \leq Isolines \leq 2047$

Returns

An integer.

Contrast Property

Specifies the current contrast value of an image.

```
(apa-get-Contrast Object)
(apa-set-Contrast Object Contrast)
```

Arguments

`Object`	The Raster object to which this property applies.
`Contrast`	An integer, in the range $0 \leq \text{Contrast} \leq 100$, specifying the contrast of the image.

Returns

An integer.

Remarks

Bi-tonal images cannot be adjusted.

You can adjust the brightness, contrast, and fade properties of an image to control its display as well as its plotted output. Doing so does not affect the original raster image file.

The brightness property darkens or lightens an image.

The contrast property can make images with poor quality easier to read.

The fade property can makes vectors easier to see over images, and can create a watermark effect in the plotted output.

ControlPoints Property

Specifies the control points of a spline.

```
(apa-get-ControlPoints Object)
(apa-set-ControlPoints Object Points)
```

Arguments

`Object`	The Spline object to which this property applies.
`Points`	A list of 3D WCS points as control points for the spline.

Returns

A list of 3D points.

Coordinate Property

Specifies the coordinates of a single vertex in the object. •

```
(apa-get-Coordinate Object Index)
(apa-set-Coordinate Object Index Point)
```

Arguments

`Object`	The 3DPoly, Leader, LightweightPolyline, Point, PolyfaceMesh, PolygonMesh, Polyline, Solid, or Trace object to which this property applies.
`Index`	The index of vertex you want to set or query.
`Point`	A point specifying the coordinates of the vertex.

Returns

A 2D point for Lightweight Polylines, a 3D point for all others.

Coordinates Property

Specifies the coordinates for each vertex in the object.

```
(apa-get-Coordinates Object)
(apa-set-Coordinates Object Coordinates)
```

Arguments

`Object`	The 3DFace, 3DPoly, Leader, LightweightPolyline, MLine, Point, PolyfaceMesh, PolygonMesh, Polyline, Solid, or Trace object to which this property applies.
`Coordinates`	A list of OCS points for Lightweight Polylines and Polylines. The Z-components are ignored.
	A list of WCS points for all others.

Returns

A list of 2D OCS points for Lightweight Polylines, 3D OCS points for Polylines, 3D and WCS points for all others.

Remarks

You can change the number of coordinates for only Lightweight Polylines and Polylines.

Count Property

Returns the number of items in the collection, dictionary, group, or selection set.

```
(apa-get-Count Object)
```

Arguments

Object   The All Collections, Block, Dictionary, Group, or SelectionSet object to which this property applies.

Returns

An integer.

CreateBackup Property

Specifies the creation use of a backup file during incremental drawing saves.

```
(apa-get-CreateBackup Object)
(apa-set-CreateBackup Object Enable)
```

Arguments

Object   The PreferencesOpenSave object to which this property applies.

Enable   non-nil to enable creation of backup files, nil to disable.

Returns

T if enabled, nil if disabled.

Remarks

This property is stored in the ISAVEBAK system variable.

CursorSize Property

Specifies the crosshairs size as a percentage of the screen size.

```
(apa-get-CursorSize Object)
(apa-set-CursorSize Object CursorSize)
```

Arguments

Object   The PreferencesDisplay object to which this property applies.

CursorSize   An integer in the range $1 <= CursorSize <= 100$ specifying the percentage of screen size for the crosshairs.

Returns

An integer.

Remarks

This property is stored in the CURSORSIZE system variable.

CustomDictionary Property

Specifies the Custom Dictionary file for spell checking.

```
(apa-get-CustomDictionary Object)
(apa-set-CustomDictionary Object FileName)
```

Arguments

Object   The PreferencesFiles object to which this property applies.

FileName   A string, specifying the custom dictionary file.

Returns

A string.

Remarks

This property is stored in the DCTCUST system variable.

CustomScale Property

Specifies the custom scale factor for the viewport.

```
(apa-get-CustomScale Object)
(apa-set-CustomScale Object CustomScale)
```

Arguments

Object	The PViewport object to which this property applies.
Scale	A positive number specifying the custom scale factor.

Returns

A real number.

Remarks

To set the viewport to a custom scale, you must first set the StandardScale property to **acVpCustomScale**. Changes will not be visible until the drawing is regenerated.

Database Property

Returns the database in which the object belongs.

```
(apa-get-Database Object)
```

Arguments

Object	The Document object to which this property applies.

Returns

The Database object that contains the *object*.

DecimalSeparator Property

Specifies the decimal separator character for decimal dimension and tolerance values.

```
(apa-get-DecimalSeparator Object)
(apa-set-DecimalSeparator Object Separator)
```

Arguments

Object	The Dim3PointAngular, DimAligned, DimAngular, DimDiametric, DimOrdinate, DimRadial, DimRotated, or Tolerance object to which this property applies.
Separator	A string , the first character of which specifies the decimal separator.

Returns

A string.

Remarks

This property overrides the value of the DIMDSEP system variable for the given dimension.

DefaultInternetURL Property

Specifies the default Internet URL (address) for the Browser command.

```
(apa-get-DefaultInternetURL Object)
(apa-set-DefaultInternetURL Object Default)
```

Arguments

Object	The PreferencesFiles object to which this property applies.
Default	A string, specifying the default URL.

Returns

A string.

Remarks

This property is stored in the INETLOCATION system variable.

DefaultOutputDevice Property

Specifies the default output device for new layouts and for model space.

```
(apa-get-DefaultOutputDevice Object)
(apa-set-DefaultOutputDevice Object Default)
```

Arguments

Object   The PreferencesOutput object to which this property applies.

Default   A string, specifying the default output device.

Returns

A string.

Remarks

This property is stored in the INETLOCATION system variable.

You can specify any system printers the system is configured for, as well as any *.pc3* files in the plotter configuration path.

If no devices are on the system, this property will return "None."

DefaultPlotStyleForLayer Property

Specifies the default plot style for Layer 0 for new drawings.

```
(apa-get-DefaultPlotStyleForLayer Object)
(apa-set-DefaultPlotStyleForLayer Object Default)
```

Arguments

Object   The PreferencesOutput object to which this property applies.

Default   A string, specifying the default plot style.

Returns

A string.

Remarks

This property is stored in the DEFLPLSTYLE system variable.

This property may be set only when PlotPolicy is set to **acPolicyNamed**.

You can enter the value "Normal" or any plot style defined in the currently loaded plot style table.

DefaultPlotStyleForObjects Property

Specifies the default plot style name for newly created objects.

```
(apa-get-DefaultPlotStyleForObjects Object)
(apa-set-DefaultPlotStyleForObjects Object Default)
```

Arguments

Object   The PreferencesOutput object to which this property applies.

Default   A string, specifying the default plot style.

Returns

A string.

Remarks

This property is stored in the DEFPLSTYLE system variable.

This property may be set only when PlotPolicy is set to **acPolicyNamed**.

You can enter the value "`Normal`" or any plot style defined in the currently loaded plot style table.

DefaultPlotStyleTable Property

Specifies the default plot style table to attach to new drawings.

```
(apa-get-DefaultPlotStyleTable Object)
(apa-set-DefaultPlotStyleTable Object Default)
```

Arguments

Object　　　　　　　The PreferencesOutput object to which this property applies.

Default　　　　　　A string, specifying the default plot style table, or "" for None.

Returns

A string.

Remarks

If you are using color-dependent plot styles, this option can be set to any color dependent plot style table found in the search path. If you are using named plot styles, this option can be set to any named plot style table.

To set color-dependent plot styles, use the PlotPolicy property.

Degree Property

Returns the degree of the spline's polynomial representation.

```
(apa-get-Degree Object)
```

Arguments

Object　　　　　　　The Spline object to which this property applies.

Returns

An integer.

Remarks

The order of a spline is the degree of the spline plus one (+1).

Delta Property

Returns the delta of a line.

```
(apa-get-Delta Object)
```

Arguments

Object　　　　　　　The line object to which this property applies.

Returns

A list of reals as in (*deltaX deltaY deltaZ*).

DemandLoadARXApp Property

Controls the demand loading of ARX applications.

```
(apa-get-DemandLoadARXApp Object)
(apa-set-DemandLoadARXApp Object Mode)
```

Arguments

Object　　　　　　　The PreferencesOpenSave object to which this property applies.

Mode　　　　　　　The sum of any combination of the following constants:

　　　　　　　　　　acDemandLoadDisabled
　　　　　　　　　　acDemandLoadOnObjectDetect
　　　　　　　　　　acDemandLoadCmdInvoke

Returns

An integer.

Remarks

This property is stored in the DEMANDLOAD system variable.

The default value for this property is `acDemandLoadOnObjectDetect` + `acDemandLoadCmdInvoke`.

`acDemandLoadOnObjectDetect`: Loads the source application when you open a drawing that contains custom objects.

`acDemandLoadCmdInvoke`: Loads the source application when you invoke one of the application's commands.

Description Property

Specifies the description of a linetype.

```
(apa-get-Description Object)
(apa-set-Description Object Description)
```

Arguments

Object	The Linetype object to which this property applies.
Description	A string.

Returns

A string.

Diameter Property

Specifies the diameter of a circle.

```
(apa-get-Diameter Object)
(apa-set-Diameter Object Diameter)
```

Arguments

Object	The Circle object to which this property applies.
Diameter	A positive number, specifying the diameter.

Returns

A real number.

Dictionaries Property

Returns the Dictionaries collection for the document.

```
(apa-get-Dictionaries Object)
```

Arguments

Object	The Document object to which this property applies.

Returns

The Dictionaries collection for the document, as an object.

DimensionLineColor Property

Specifies the color of dimension line arrowheads and leader line arrowheads.

```
(apa-get-DimensionLineColor Object)
(apa-set-DimensionLineColor Object Color)
```

Arguments

Object	The Dim3PointAngular, DimAligned, DimAngular, DimDiametric, DimRadial , DimRotated, Leader, or Tolerance object to which this property applies.

Color	An ACI integer in the range $0 \leq Color \leq 256$, or one of the following:
	`acByBlock`
	`acByLayer`

```
                        acRed
                        acYellow
                        acGreen
                        acCyan
                        acBlue
                        acMagenta
                        acWhite
```

Returns

An Integer.

Remarks

This property overrides the value of the DIMCLRD system variable for the given dimension.

DimensionLineExtend Property

Specifies the distance to extend the dimension line beyond the extension line.

```
(apa-get-DimensionLineExtend Object)
(apa-set-DimensionLineExtend Object Extend)
```

Arguments

Object	The Dim3PointAngular, DimAligned, DimAngular, DimDiametric, DimRadial, DimRotated, Leader, or Tolerance object to which this property applies.
Extend	A positive number specifying the distance to extend the dimension line beyond the extension line.

Returns

A real number.

Remarks

This property overrides the value of the DIMDLE system variable for the given dimension.

DimensionLineWeight Property

Specifies the line weight for dimension lines.

```
(apa-get-DimensionLineWeight Object)
(apa-set-DimensionLineWeight Object LineWeight)
```

Arguments

Object	The Dim3PointAngular, DimAligned, DimAngular, DimDiametric, DimOrdinate, DimRadial, or DimRotated object to which this property applies.
LineWeight	One of the following:

```
                        acLnWtByLayer
                        acLnWtByBlock
                        acLnWtByLwDefault
                        acLnWt000
                        acLnWt005
                        acLnWt009
                        acLnWt013
                        acLnWt015
                        acLnWt018
                        acLnWt020
                        acLnWt025
                        acLnWt030
                        acLnWt035
                        acLnWt040
                        acLnWt050
                        acLnWt053
```

```
acLnWt060
acLnWt070
acLnWt080
acLnWt090
acLnWt100
acLnWt106
acLnWt120
acLnWt140
acLnWt158
acLnWt200
acLnWt211
```

Returns

An integer.

Remarks

This property overrides the value of the DIMLWD system variable for the given dimension.
LineWeight settings are visible only when the LineWeightDisplay property is set to T.

DimLine1Suppress Property

Specifies the suppression of the first dimension line.

```
(apa-get-DimLine1Suppress Object)
(apa-set-DimLine1Suppress Object Suppress)
```

Arguments

Object   The Dim3PointAngular, DimAligned, DimAngular, DimDiametric, or DimRotated
                    object to which this property applies.

Suppress   non-nil to suppress, nil to not suppress.

Returns

T if suppressed, nil if not.

Remarks

This property overrides the value of the DIMSD1 system variable for the given dimension.

DimLine2Suppress Property

Specifies the suppression of the first dimension line.

```
(apa-get-DimLine2Suppress Object)
(apa-set-DimLine2Suppress Object Suppress)
```

Arguments

Object   The Dim3PointAngular, DimAligned, DimAngular, DimDiametric, or DimRotated
                    object to which this property applies.

Suppress   non-nil to suppress, nil to not suppress.

Returns

T if suppressed, nil if not.

Remarks

This property overrides the value of the DIMSD2 system variable for the given dimension.

DimLineInside Property

Suppresses drawing of dimension lines outside the extension lines.

```
(apa-get-DimLineInside Object)
(apa-set-DimLineInside Object Suppress)
```

Arguments

Object The Dim3PointAngular, DimAligned, DimAngular, or DimRotated object to which this property applies.

Suppress non-nil to suppresses, nil to not suppress dimension lines outside the extension lines.

Returns

T if suppressed, nil if not.

Remarks

This property overrides the value of the DIMSOXD system variable for the given dimension.

DimLineSuppress Property

Specifies the suppression of the dimension line for radial dimensions.

```
(apa-get-DimLineSuppress Object)
(apa-set-DimLineSuppress Object Suppress)
```

Arguments

Object The DimRadial object to which this property applies.

Suppress non-nil to suppress, nil to not suppress.

Returns

T if suppressed, nil if not.

Remarks

This property overrides the value of the DIMSD2 system variable for the given dimension.

Dimstyles Property

Returns the Dimstyles collection for the document.

```
(apa-get-Dimstyles Object)
```

Arguments

Object The Document object to which this property applies.

Returns

The Dimstyles collection for the document, as an object.

Direction Property

Specifies the viewing direction for a 3D visualization of the drawing.

```
(apa-get-Direction Object)
(apa-set-Direction Object Direction)
```

Arguments

Object The View, Viewport, or PViewport object to which this property applies.

Direction A Vector.

Returns

A Vector

Remarks

This property puts the viewer in a position as if looking back at the origin (0, 0, 0) from a specified point in space. Once you've changed the direction of a viewport, you must reset the active viewport.

DirectionVector Property

Specifies the direction for the ray, tolerance, or xline through a vector.

```
(apa-get-DirectionVector Object)
```

`(apa-set-DirectionVector `*`Object Direction`*`)`

Arguments

Object   The Ray, Tolerance, or XLine object to which this property applies.

Direction   A Vector specifying the direction of the object.

Returns

A Vector

Display Property

Returns the PreferencesDisplay object.

`(apa-get-Display `*`Object`*`)`

Arguments

Object   The Preferences object to which this property applies.

Returns

The PreferencesDisplay object.

Remarks

There are two types of options on the Display tab of the Options dialog box:

- Those that are stored in the current drawing, affecting only that drawing. These are held in the DatabasePreferences object.

- Those that are stored in the system registry, affecting the AutoCAD application. These are held in the PreferencesDisplay object.

The following options from the Display tab are stored in the DatabasePreferences object: **ContourLinesPerSurface**, **DisplaySilhouette**, **RenderSmoothness**, **SegmentPerPolyline**, **SolidFill**, and **TextFrameDisplay**.

DisplayGrips Property

Controls the display of selection set grips within blocks.

`(apa-get-DisplayGrips `*`Object`*`)`
`(apa-set-DisplayGrips `*`Object Display`*`)`

Arguments

Object   The PreferencesSelection object to which this property applies.

Display   non-nil to display grips, nil to not display grips.

Returns

T if displayed, nil if not.

Remarks

This property is stored in the GRIPS system variable.

DisplayGripsWithinBlocks Property

Controls the display of selection set grips.

`(apa-get-DisplayGripsWithinBlocks `*`Object`*`)`
`(apa-set-DisplayGripsWithinBlocks `*`Object Display`*`)`

Arguments

Object   The PreferencesSelection object to which this property applies.

Display   non-nil to display grips, nil to not display grips.

Returns

T if displayed, nil if not.

Remarks

This property is stored in the GRIPBLOCK system variable.

DisplayLayoutTabs Property

Controls the display of Model and Layout tabs in the drawing editor.

```
(apa-get-DisplayLayoutTabs Object)
(apa-set-DisplayLayoutTabs Object Display)
```

Arguments

Object	The PreferencesDisplay object to which this property applies.
Display	non-nil to display layout tabs, nil to not display grips.

Returns

T if displayed, nil if not.

DisplayLocked Property

Specifies if the viewport is locked.

```
(apa-get-DisplayLocked Object)
(apa-set-DisplayLocked Object Locked)
```

Arguments

Object	The PViewport object to which this property applies.
Locked	non-nil to lock the viewport, nil to unlock it.

Returns

T if locked, nil if not.

Remarks

If the active viewport is locked, AutoCAD switches to paper space before zooming or panning, and returns to model space when done.

DisplayOLEScale Property

Controls the display of the OLE scaling dialog when OLE objects are inserted into a drawing.

```
(apa-get-DisplayOLEScale Object)
(apa-set-DisplayOLEScale Object Display)
```

Arguments

Object	The PreferencesSystem object to which this property applies.
Display	non-nil to display the dialog, nil to not display the dialog.

Returns

T if displayed, nil if not.

DisplayScreenMenu Property

Controls the display of the AutoCAD screen menu.

```
(apa-get-DisplayScreenMenu Object)
(apa-set-DisplayScreenMenu Object Display)
```

Arguments

Object	The PreferencesDisplay object to which this property applies.
Display	non-nil to display the menu, nil to not display the menu.

Returns

T if displayed, nil if not.

DisplayScrollBars Property

Controls the display of scroll bars in drawing window.

```
(apa-get-DisplayScrollBars Object)
(apa-set-DisplayScrollBars Object DisplayScrollBars)
```

Arguments

Object   The PreferencesDisplay object to which this property applies.

Display   non-`nil` to display the menu, `nil` to not display the menu.

Returns

`T` if displayed, `nil` if not.

DisplaySilhouette Property

Controls the display of silhouette curves of solid objects.

```
(apa-get-DisplaySilhouette Object)
(apa-set-DisplaySilhouette Object Display)
```

Arguments

Object   The DatabasePreferences object to which this property applies.

Display   non-`nil` to display the silhouette curves, `nil` to not.

Returns

`T` if displayed, `nil` if not.

Remarks

This property is stored in the DISPSILH system variable.

DockedVisibleLines Property

Specifies the number of lines of text to display in the command window, when said window is docked.

```
(apa-get-DockedVisibleLines Object)
(apa-set-DockedVisibleLines Object Lines)
```

Arguments

Object   The PreferencesDisplay object to which this property applies.

Lines   A number.

Returns

An integer

DockStatus Property

Determines if the toolbar is docked or floating.

```
(apa-get-DockStatus Object)
```

Arguments

Object   The Toolbar object to which this property applies.

Returns

An integer   One of the following:

                      `acToolbarDockTop`
                      `acToolbarDockBottom`
                      `acToolbarDockLeft`
                      `acToolbarDockRight`
                      `acToolbarFloating`

Document Property

Retrieves the document (drawing) to which an object belongs.

(apa-get-Document *Object*)

Arguments

Object   The All Drawing objects, Block, Blocks, Dictionary, Dictionaries, DimStyle, Dimstyles, Group, Groups, Layer, Layers, Layout, Layouts, Linetype, Linetypes, ModelSpace Collection, PaperSpace Collection, PlotConfiguration, PlotConfigurations, RegisteredApplication, RegisteredApplications, TextStyle, TextStyles, UCS, UCSs, View, Views, Viewport,, Viewports, or XRecord object to which this property applies.

Returns

The Document object to which the specified object belongs.

Documents Property

Returns the Documents collection for the application.

(apa-get-Documents *Object*)

Arguments

Object   The Application object to which this property applies.

Returns

The Documents collection for the document, as an object.

Drafting Property

Returns the PreferencesDrafting object.

(apa-get-Drafting *Object*)

Arguments

Object   The Preferences object to which this property applies.

Returns

The PreferencesDrafting object.

DrawingDirection Property

Specifies the direction in which the MText paragraph is to be drawn.

(apa-get-DrawingDirection *Object*)
(apa-set-DrawingDirection *Object Direction*)

Arguments

Object   The MText object to which this property applies.

Direction   An integer. One of the following:

acLeftToRight
acTopToBottom
acRightToLeft   Reserved for future use
acBottomToTop   Reserved for future use
acByStyle   Reserved for future use

Returns

An integer.

Remarks

For some languages, text is read horizontally, left to right. For others, text is read vertically, top to bottom.

The settings **acRightToLeft**, **acBottomToTop**, and **acByStyle** are reserved for future use and cannot be used in this release.

DriversPath Property

Specifies the directories AutoCAD searches for ADI device drivers.

```
(apa-get-DriversPath Object)
(apa-set-DriversPath Object Path)
```

Arguments

Object   The PreferencesFiles object to which this property applies.

Path   A string, specifying the directories in which to search.

Returns

A string.

Remarks

Semicolons separate multiple directories.

Elevation Property

Specifies the elevation of the hatch or polyline.

```
(apa-get-Elevation Object)
(apa-set-Elevation Object Elevation)
```

Arguments

Object   The Hatch, LightweightPolyline, or Polyline object to which this property applies.

Elevation   A number specifying the elevation of the object.

Returns

A real number.

ElevationModelSpace Property

Specifies the current elevation setting in the model space.

```
(apa-get-ElevationModelSpace Object)
(apa-set-ElevationModelSpace Object Elevation)
```

Arguments

Object   The Document object to which this property applies.

Elevation   A number specifying the Z value that is used whenever a 3D point is expected from the user, but a 2D point is entered.

Returns

A real number.

Remarks

The current elevation is maintained separately in model space and paper space.

ElevationPaperSpace Property

Specifies the current elevation setting in the paper space.

```
(apa-get-ElevationPaperSpace Object)
(apa-set-ElevationPaperSpace Object Elevation)
```

Arguments

Object   The Document object to which this property applies.

Elevation   A number specifying the Z value that is used whenever a 3D point is expected from the user, but a 2D point is entered.

Returns

A real number.

Remarks

The current elevation is maintained separately in model space and paper space.

Enable Property

Enables the specified popup menu item.

```
(apa-get-Enable Object)
(apa-set-Enable Object Enable)
```

Arguments

object   The PopupMenuItem object to which this property applies.

Enable   non-nil to enable the menu item, nil to disable it.

Returns

T if the menu item is enabled, nil it isn't.

EnableStartupDialog Property

Enables the AutoCAD Startup Dialog.

```
(apa-get-EnableStartupDialog Object)
(apa-set-EnableStartupDialog Object Enable)
```

Arguments

object   The PreferencesSystem object to which this property applies.

Enable   non-nil to enable the startup dialog, nil to disable it

Returns

T if the startup dialog is enabled, nil it isn't.

EndAngle Property

Specifies the End Angle of an arc, or ellipse.

```
(apa-get-EndAngle Object)
(apa-set-EndAngle Object EndAngle)
```

Arguments

Object   The Arc or Ellipse object to which this method applies.

EndAngle   A number specifying the End Angle in radians. Use 2p to specify a closed ellipse.

Returns

A real number.

EndParameter Property

Specifies the end parameter for an ellipse.

```
(apa-get-EndParameter Object)
(apa-set-EndParameter Object EndParameter)
```

Arguments

Object   The Ellipse object to which this method applies.

EndParameter   A number specifying the End Parameter.

Returns

A real number.

Remarks

The start and end parameters of the ellipse are supposedly computed with on the following equation:

$$P(\theta) = A \times \cos(\theta) + B \times \sin(\theta)$$

where A and B are the semi-major and semi-minor axes respectively, and θ is the starting or ending angle of the ellipse.

In fact, this property, as set and returned by AutoCAD, seems to have the same value as the EndAngle property.

EndPoint Property

Specifies the endpoint for an arc, line, or ellipse.

```
(apa-get-EndPoint Object)
(apa-set-EndPoint Object EndPoint)
```

Arguments

Object   The Arc, Line, or Ellipse object to which this method applies.

EndPoint   A point. You may not set the endpoint for an ellipse or an arc.

Returns

A 3D point.

EndTangent Property

Specifies the end tangent of the spline as a directional vector.

```
(apa-get-EndTangent Object)
(apa-set-EndTangent Object EndTangent)
```

Arguments

Object   The Spline object to which this method applies.

EndTangent   Vector specifying the tangency of the curve at the end of the spline.

Returns

A 3D vector.

ExtensionLineColor Property

Specifies the color of dimension extension lines.

```
(apa-get-ExtensionLineColor Object)
(apa-set-ExtensionLineColor Object Color)
```

Arguments

Object   The Dim3PointAngular, DimAligned, DimAngular, DimOrdinate, or DimRotated
                object to which this property applies.

Color   An ACI integer in the range $0 \leq Color \leq 256$, or one of the following:

                acByBlock
                acByLayer
                acRed
                acYellow
                acGreen
                acCyan
                acBlue
                acMagenta
                acWhite

Returns

An Integer.

Remarks

This property overrides the value of the DIMCLRE system variable for the given dimension.

ExtensionLineExtend Property

Specifies the distance to extend the dimension line beyond the extension line.

```
(apa-get-ExtensionLineExtend Object)
(apa-set-ExtensionLineExtend Object Extend)
```

Arguments

Object
: The Dim3PointAngular, DimAligned, DimAngular, DimDiametric, DimRadial, DimRotated, Leader, or Tolerance object to which this property applies.

Extend
: A positive number specifying the distance to extend the dimension line beyond the extension line.

Returns

A real number.

Remarks

This property overrides the value of the DIMEXE system variable for the given dimension.

ExtensionLineOffset Property

Specifies the distance to offset the extension lines from their origins.

```
(apa-get-ExtensionLineOffset Object)
(apa-set-ExtensionLineOffset Object Offset)
```

Arguments

Object
: The Dim3PointAngular, DimAligned, DimAngular, DimDiametric, DimRadial, DimRotated, Leader, or Tolerance object to which this property applies.

Offset
: A positive number specifying the distance to offset the extension lines from their origins.

Returns

A real number.

Remarks

This property overrides the value of the DIMEXO system variable for the given dimension.

ExtensionLineWeight Property

Specifies the line weight for extension lines.

```
(apa-get-ExtensionLineWeight Object)
(apa-set-ExtensionLineWeight Object LineWeight)
```

Arguments

Object
: The Dim3PointAngular, DimAligned, DimAngular, DimDiametric, DimOrdinate, DimRadial, or DimRotated object to which this property applies.

Lineweight
: One of the following:

```
acLnWtByLayer
acLnWtByBlock
acLnWtByLwDefault
acLnWt000
acLnWt005
acLnWt009
acLnWt013
acLnWt015
acLnWt018
acLnWt020
acLnWt025
acLnWt030
acLnWt035
acLnWt040
acLnWt050
acLnWt053
acLnWt060
acLnWt070
```

```
acLnWt080
acLnWt090
acLnWt100
acLnWt106
acLnWt120
acLnWt140
acLnWt158
acLnWt200
acLnWt211
```

Returns

An integer.

Remarks

This property overrides the value of the DIMLWE system variable for the given extension.

ExtLine1EndPoint Property

Specifies the endpoint of the first extension line for Angular dimensions.

```
(apa-get-ExtLine1EndPoint Object)
(apa-set-ExtLine1EndPoint Object EndPoint)
```

Arguments

Object	The Dim3PointAngular or DimAngular object to which this method applies.
EndPoint	A point specifying the endpoint of the first extension line.

Returns

A 3D point.

ExtLine1Point Property

Specifies the origin of extension line 1 for an Aligned dimension.

```
(apa-get-ExtLine1Point Object)
(apa-set-ExtLine1Point Object Point)
```

Arguments

Object	The DimAligned object to which this method applies.
Point	A point specifying the origin of the first extension line.

Returns

A 3D point.

ExtLine1StartPoint Property

Specifies the origin of extension line 1 for Angular dimensions.

```
(apa-get-ExtLine1StartPoint Object)
(apa-set-ExtLine1StartPoint Object Point)
```

Arguments

Object	The DimAngular object to which this method applies.
Point	A point specifying the origin of the first extension line.

Returns

A 3D point.

Remarks

The ExtLine1StartPoint and ExtLine2StartPoint properties correspond to the *FirstEndPoint* and *SecondEndPoint* parameters of the AddDimAngular method.

ExtLine1Suppress Property

Specifies the suppression of the first extension line.

```
(apa-get-ExtLine1Suppress Object)
(apa-set-ExtLine1Suppress Object Suppress)
```

Arguments

Object	The Dim3PointAngular, DimAligned, DimAngular, or DimRotated object to which this property applies.
Suppress	non-nil to suppress, nil to not suppress.

Returns

T if suppressed, nil if not.

Remarks

This property overrides the value of the DIMSE1 system variable for the given dimension.

ExtLine2EndPoint Property

Specifies the endpoint of the second extension line for Angular dimensions.

```
(apa-get-ExtLine2EndPoint Object)
(apa-set-ExtLine2EndPoint Object EndPoint)
```

Arguments

Object	The Dim3PointAngular or DimAngular object to which this method applies.
EndPoint	A point specifying the endpoint of the first extension line.

Returns

A 3D point.

ExtLine2Point Property

Specifies the origin of extension line 2 for an Aligned dimension.

```
(apa-get-ExtLine2Point Object)
(apa-set-ExtLine2Point Object Point)
```

Arguments

Object	The DimAligned object to which this method applies.
Point	A point specifying the origin of the first extension line.

Returns

A 3D point.

ExtLine2StartPoint Property

Specifies the origin of extension line 2 for Angular dimensions.

```
(apa-get-ExtLine2StartPoint Object)
(apa-set-ExtLine2StartPoint Object Point)
```

Arguments

Object	The DimAngular object to which this method applies.
Point	A point specifying the origin of the second extension line.

Returns

A 3D point.

Remarks

The ExtLine2StartPoint and ExtLine2StartPoint properties correspond to the *FirstEndPoint* and *SecondEndPoint* parameters of the AddDimAngular method.

ExtLine2Suppress Property

Specifies the suppression of the first extension line.

```
(apa-get-ExtLine2Suppress Object)
(apa-set-ExtLine2Suppress Object Suppress)
```

Arguments

Object	The Dim3PointAngular, DimAligned, DimAngular, or DimRotated object to which this property applies.
Suppress	non-nil to suppress, nil to not suppress.

Returns

T if suppressed, nil if not.

Remarks

This property overrides the value of the DIMSE2 system variable for the given dimension.

Fade Property

Specifies the current fade value of an image.

```
(apa-get-Fade Object)
(apa-set-Fade Object Fade)
```

Arguments

Object	The Raster object to which this property applies.
Fade	An integer in the range $0 \leq Fade \leq 100$, specifying the fade of the image.

Returns

An integer.

Remarks

Bi-tonal images cannot be adjusted

You can adjust the brightness, contrast, and fade properties of an image to control its display as well as its plotted output. Doing so does not affect the original raster image file.

The brightness property darkens or lightens an image.

The contrast property can make images with poor quality easier to read.

The fade property can makes vectors easier to see over images, and can create a watermark effect in the plotted output.

FieldLength Property

Specifies the field length of the attribute.

```
(apa-get-FieldLength Object)
(apa-set-FieldLength Object FieldLength)
```

Arguments

Object	The Attribute or AttributeRef object to which this property applies.
FieldLength	An positive integer.

Returns

An integer.

Remarks

It's not clear to me what, if any, use AutoCAD makes of the FieldLength property.

Files Property

Returns the PreferencesFiles object.

```
(apa-get-Files Object)
```

Arguments

`Object`   The Preferences object to which this property applies.

Returns

The PreferencesFiles object.

Remarks

All the options on the Files tab of the Options dialog box are stored in the system registry, affecting the AutoCAD application. These are held in the PreferencesFiles object.

Fit Property

Specifies the placement of text and arrowheads inside or outside extension lines

```
(apa-get-Fit Object)
(apa-set-Fit Object Fit)
```

Arguments

`Object`   The Dim3PointAngular, DimAligned, DimAngular, DimDiametric, DimRadial, or DimRotated object to which this property applies.

`Fit`   An integer specifying which get moved outside the dimension lines where there is not sufficient room to place both text and arrows inside the extension lines. One of the following:

`acTextAndArrows`	Moved Simultaneously
`acArrowsOnly`	Arrows are moved first
`acTextOnly`	Text is moved first
`acBestFit`	Either text or arrows, for the best fit.

Returns

An integer.

Remarks

This property overrides the value of the DIMATFIT system variable for the given dimension.

FitPoints Property

Specifies the fit points of a spline.

```
(apa-get-FitPoints Object)
(apa-set-FitPoints Object Points)
```

Arguments

`Object`   The Spline object to which this property applies.

`Points`   A list of 3D WCS points as fit points for the spline.

Returns

A list of 3D points.

FitTolerance Property

Refits the spline to the existing points with new tolerance values.

```
(apa-get-FitTolerance Object)
(apa-set-FitTolerance Object Tolerance)
```

Arguments

`Object`   The Spline object to which this property applies.

`Tolerance`   A positive number specifying the fit tolerance for the spline.

Returns

A real number.

FloatingRows Property

Specifies the number of rows for a floating toolbar.

```
(apa-get-FloatingRows Object)
(apa-Set-FloatingRows Object Rows)
```

Arguments

Object   The Toolbar object to which this property applies.

Rows   A number specifying the number of rows for the floating toolbar.

Returns

An integer.

Flyout Property

Returns the toolbar associated with a flyout toolbar item.

```
(apa-get-Flyout Object)
```

Arguments

Object   The ToolbarItem object to which this property applies.

Returns

The toolbar object that contains the flyout.

```
(setq mgObj (apa-Item apMenuGroups "ACAD"))
(setq tbarsObj (apa-get-Toolbars mgObj))
(setq tbarObj (apa-Item tbarsObj "Standard Toolbar"))
(setq tbaritemObj (apa-Item tbarObj "Object Snap"))
(apa-get-Flyout tbaritemObj)
```

FontFile Property

Specifies the name of the font file associated with the text style.

```
(apa-get-FontFile Object)
(apa-set-FontFile Object FileName)
```

Arguments

Object   The TextStyle object to which this property applies.

FileName   String, specifying the name and extension the font file.

Returns

A string.

Remarks

You must call the Regen method to see the changes to the text.

FontFileMap Property

Specifies the location of the font-mapping file.

```
(apa-get-FontFileMap Object)
(apa-set-FontFileMap Object FileName)
```

Arguments

Object   The PreferencesFiles object to which this property applies.

FileName   String, specifying the name and extension the font mapping (*.fmp*) file.

Returns

A string.

ForceLineInside Property

Forces a dimension line be drawn between the extension lines even when the text is placed outside.

```
(apa-get-ForceLineInside Object)
(apa-set-ForceLineInside Object Inside)
```

Arguments

Object	The Dim3PointAngular, DimAligned, DimAngular, DimDiametric, DimRadial, or DimRotated object to which this property applies.
Inside	non-nil to force dimension lines inside, nil to not.

Returns

T if forced inside, nil if not.

Remarks

This property overrides the value of the DIMTOFL system variable for the given dimension.

For radius and diameter dimensions, draws a dimension line inside the circle or arc.

FractionFormat Property

Specifies the format of fractions in dimensions and tolerances.

```
(apa-get-FractionFormat Object)
(apa-set-FractionFormat Object Format)
```

Arguments

Object	The Dim3PointAngular, DimAligned, DimAngular, DimDiametric, DimOrdinate, DimRadial, DimRotated, or Tolerance object to which this property applies.
Format	An integer specifying the format. One of the following:

        acHorizontal
        acDiagonal
        acNotStacked.

Returns

An integer.

Remarks

This property overrides the value of the DIMFRAC system variable for the given dimension.

Freeze Property

Specifies the freeze status of a layer.

```
(apa-get-Freeze Object)
(apa-set-Freeze Object Freeze)
```

Arguments

Object	The Layer object to which this property applies.
Freeze	non-nil to freeze the layer, nil to thaw it.

Returns

T if frozen, nil if thawed.

FullCRCValidation Property

Specifies if full time CRC validation should be enabled.

```
(apa-get-FullCRCValidation Object)
(apa-set-FullCRCValidation Object Enable)
```

Arguments

Object   The PreferencesOpenSave object to which this property applies.

Enable   non-nil to enable; nil to disable.

Returns

T if enabled, nil if disabled.

Remarks

If your drawings are being corrupted, set FullCRCValidation to T. This will determine if they're being corrupted while running AutoCAD.

FullName Property

Returns the fully qualified path name of the application or document.

```
(apa-get-FullName Object)
```

Arguments

Object   The Application or Document object to which this property applies.

Returns

A string.

FullScreenTrackingVector Property

Controls the display of full screen tracking vectors.

```
(apa-get-FullScreenTrackingVector Object)
(apa-set-FullScreenTrackingVector Object Enable)
```

Arguments

Object   The PreferencesDrafting object to which this property applies.

Enable   non-nil to enable; nil to disable.

Returns

T if enabled, nil if disabled.

Remarks

This property is stored in the TRACKPATH system variable.

GraphicsWinLayoutBackgrndColor Property

Specifies the background color for the paper space layouts.

```
(apa-get-GraphicsWinLayoutBackgrndColor Object)
(apa-set-GraphicsWinLayoutBackgrndColor Object Color)
```

Arguments

Object   The PreferencesDisplay object to which this method applies.

Color   An integer specifying an OLE_COLOR. One of the following:

                        vbBlack
                        vbRed
                        vbYellow
                        vbGreen
                        vbCyan
                        vbBlue
                        vbMagenta
                        vbWhite

Returns

An integer.

GraphicsWinModelBackgrndColor Property

Specifies the background color for the model space window.

```
(apa-get-GraphicsWinModelBackgrndColor Object)
(apa-set-GraphicsWinModelBackgrndColor Object Color)
```

Arguments

Object	The PreferencesDisplay object to which this method applies.
Color	An integer specifying an OLE_COLOR. One of the following:

> **vbBlack**
> **vbRed**
> **vbYellow**
> **vbGreen**
> **vbCyan**
> **vbBlue**
> **vbMagenta**
> **vbWhite**

Returns

An integer.

GridOn Property

```
(apa-get-GridOn Object)
(apa-set-GridOn Object Enable)
```

Arguments

Object	The Viewport or PViewport object to which this property applies.
Enable	non-nil to enable; nil to disable.

Returns

T if enabled, nil if disabled.

Remarks

This property is stored in the GRIDMODE system variable.

Once you've changed the GridOn property of a viewport, you must reset the active viewport to see the changes.

GripColorSelected Property

Specifies the color of selected (hot) grips.

```
(apa-get-GripColorSelected Object)
(apa-set-GripColorSelected Object Color)
```

Arguments

Object	The PreferencesSelection object to which this property applies.
Color	An integer specifying the ACI of selected grips. One of the following:

> **acRed**
> **acYellow**
> **acGreen**
> **acCyan**
> **acBlue**
> **acMagenta**
> **acWhite**

Returns

An Integer.

Remarks

This property is stored in the GRIPHOT system variable.

GripColorUnselected Property

Specifies the color of unselected (warm or cold) grips.

```
(apa-get-GripColorUnselected Object)
(apa-set-GripColorUnselected Object Color)
```

Arguments

Object	The PreferencesSelection object to which this property applies.
Color	An integer specifying the ACI of unselected grips. One of the following:

```
acRed
acYellow
acGreen
acCyan
acBlue
acMagenta
acWhite
```

Returns

An Integer.

Remarks

This property is stored in the GRIPCOLOR system variable.

GripSize Property

Specifies the size of grips.

```
(apa-get-GripSize Object)
(apa-set-GripSize Object Size)
```

Arguments

Object	The PreferencesSelection object to which this property applies.
Size	An integer specifying the size, in pixels, of the box to display the grips.

Returns

An Integer.

Remarks

This value of this property is stored in the GRIPSIZE system variable.

Groups Property

Returns the Groups collection for the document.

```
(apa-get-Groups Object)
```

Arguments

Object	The Document object to which this property applies.

Returns

The Groups collection for the document, as an object.

HWND Property

Returns the window handle of the document window frame.

```
(apa-get-HWND Object)
```

Arguments

Object   The Document object to which this property applies.

Returns

An integer.

Remarks

This handle (Hwnd) of a document window can be used with WindowsAPI calls, or ActiveX components that require a handle to a window.

Handle Property

Returns the handle of an object.

(apa-get-Handle *Object*)

Arguments

Object   The All Drawing objects, Block, Blocks, Dictionary, Dictionaries, DimStyle, Dimstyles, Group, Groups, Layer, Layers, Layout, Layouts, Linetype, Linetypes, ModelSpace Collection, PaperSpace Collection, PlotConfiguration, PlotConfigurations, RegisteredApplication, RegisteredApplications, TextStyle, TextStyles, UCS, UCSs, View, Views, Viewport, Viewports or XRecord object to which this property applies.

Returns

A string.

HasAttributes Property

Determines if the block reference has attributes attached to it.

(apa-get-HasAttributes *Object*)

Arguments

Object   The BlockRef, ExternalReference, or MInsertBlock object to which this property applies.

Returns

T if it does, nil if it doesn't.

HatchStyle Property

Specifies the hatch style.

(apa-get-HatchStyle *Object*)
(apa-set-HatchStyle *Object Style*)

Arguments

Object   The Hatch object to which this property applies.
Style   An integer. One of the following:

acHatchStyleNormal
acHatchStyleOuter
acHatchStyleIgnore

Returns

An integer.

Height Property

Specifies the Height of the main application window, attribute, raster, shape, text, toolbar, view, or viewport.

(apa-get-Height *Object*)
(apa-set-Height *Object Height*)

Arguments

Object   The Application, Attribute, AttributeRef, MText, PViewport, Raster, Shape, Text, TextStyle, Toolbar, Viewport, or View object to which this property applies.

Height   A positive number.
In pixels, for Application, Raster and Toolbar objects.
In current drawing units for all others.

Returns

An integer, in pixels, in pixels, for Application, Raster, and Toolbar objects.
A real in current drawing units for all others.

HelpFilePath Property

Specifies the location of the AutoCAD Help file.

```
(apa-get-HelpFilePath Object)
(apa-set-HelpFilePath Object FileName)
```

Arguments

Object   The PreferencesFiles object to which this property applies.

FileName   A string, specifying the name and location of the AutoCAD Help File.

Returns

A string.

HelpString Property

Specifies the help string for the toolbar, toolbar item, or menu item.

```
(apa-get-HelpString Object)
(apa-set-HelpString Object HelpString)
```

Arguments

Object   The PreferencesFiles object to which this property applies.

HelpString   The text string to appears in the AutoCAD status line when a user highlights the toolbar, toolbar item, or menu item.

Returns

A string.

HistoryLines Property

Specifies the number of lines of text to retain in the text window.

```
(apa-get-HistoryLines Object)
(apa-set-HistoryLines Object Lines)
```

Arguments

Object   The PreferencesDisplay object to which this property applies.

Lines   A number.

Returns

An integer.

HorizontalTextPosition Property

Specifies the horizontal alignment for dimension text.

```
(apa-get-HorizontalTextPosition Object)
(apa-set-HorizontalTextPosition Object Alignment)
```

Arguments

Object   The Dim3PointAngular, DimAligned, DimAngular, DimDiametric, DimRadial, or
                          DimRotated object to which this property applies.

Alignment   An integer specifying the alignment. One of the following:

                          acHorzCentered
                          acFirstExtensionLine
                          acSecondExtensionLine
                          acOverFirstExtension
                          acOverSecondExtension

Returns

The alignment as an integer.

Remarks

This property overrides the value of the DIMJUST system variable for the given dimension.

ImageFile Property

Specifies the full path and file name of the raster image file.

```
(apa-get-ImageFile Object)
(apa-set-ImageFile Object FileName)
```

Arguments

Object   The Raster object to which this method applies.

FileName   A string, specifying the path and file name of the image.

Returns

A string.

Remarks

Searches the AutoCAD Search path.

This property is similar to the Name property, except this property *may* contain the path information and the Name property does not.

You can use this property to load a new raster image into an existing raster entity.

ImageFrameHighlight Property

Controls the display of raster images during selection.

```
(apa-get-ImageFrameHighlight Object)
(apa-set-ImageFrameHighlight Object Highlight)
```

Arguments

Object   The PreferencesDisplay object to which this method applies.

Highlight   non-nil to highlight only the frame when images are selected. nil to highlight both
                          the image and the frame.

Returns

T if only the frame is highlighted, nil if both the image and the frame are highlighted.

Remarks

This property is stored in the IMAGEHLT system variable.

ImageHeight Property

Specifies the Height of the raster image in current drawing units.

```
(apa-get-ImageHeight Object)
(apa-set-ImageHeight Object Height)
```

Arguments

Object	The Raster object to which this property applies.
Height	A positive number, specifying the height in current drawing units.

Returns

A real number.

ImageVisibility Property

Specifies the Visibility of the raster image.

```
(apa-get-ImageVisibility Object)
(apa-set-ImageVisibility Object Visibility)
```

Arguments

Object	The Raster object to which this property applies.
Visibility	non-nil specifies the object is visible, nil that it isn't.

Returns

T if the object is visible, nil that it isn't

ImageWidth Property

Specifies the Width of the raster image in current drawing units.

```
(apa-get-ImageWidth Object)
(apa-set-ImageWidth Object Width)
```

Arguments

Object	The raster object to which this property applies.
Width	A positive number, specifying the width in current drawing units.

Returns

A real number.

IncrementalSavePercent Property

Specifies the amount of wasted space tolerated in a drawing file.

```
(apa-get-IncrementalSavePercent Object)
(apa-set-IncrementalSavePercent Object Percent)
```

Arguments

Object	The PreferencesOpenSave object to which this property applies.
Percent	An number in the range $0 \leq Percent \leq 100$.
	When the estimate of wasted space within the file exceeds *Percent* percent, the next save will be a full save, resetting the wasted space to 0.
	If *Percent* is set to 0, every save is a full save.

Returns

An Integer.

Remarks

This property is stored in the ISAVEPERCENT system variable.

Index Property

Specifies the index of the menu or toolbar item.

```
(apa-get-Index Object)
```

Arguments

Object	The PopupMenuItem or ToolbarItem object to which this property applies.

Returns

An integer specifying the position of the menu or toolbar item. The first position in the index is 0.

InsertionPoint Property

Specifies the Insertion point for a tolerance, text, block, or shape.

```
(apa-get-InsertionPoint Object)
(apa-set-InsertionPoint Object InsertionPoint)
```

Arguments

Object	The Attribute, AttributeRef, BlockRef, ExternalReference, MInsertBlock, MText, Shape, Text, or Tolerance Symbol object to which this property applies.
InsertionPoint	A point in the WCS specifying the insertion point.

Returns

A 3D point.

Remarks

Text aligned to **acAlignmentLeft** uses only the InsertPoint property to locate the text.

You cannot set the TextAlignmentPoint property while the text is aligned to **acAlignmentLeft**.

Text aligned to **acAlignmentAligned**, or **acAlignmentFit** uses both the InsertPoint and TextAlignmentPoint properties.

All other alignments use only the TextAlignmentPoint property to locate the text.

Invisible Property

Specifies if the attribute or attribute reference is invisible.

```
(apa-get-Invisible Object)
(apa-set-Invisible Object Visibility)
```

Arguments

Object	The Attribute or AttributeReference object to which this property applies.
Visibility	non-nil specifies the object is visible, nil that it isn't.

Returns

T if the object is visible, nil that it isn't.

IsLayout Property

Determines if the given block is a layout block.

```
(apa-get-IsLayout Object)
```

Arguments

Object	The Block object to which this property applies.

Returns

T if the object is a layout block, nil that it isn't.

ISOPenWidth Property

Specifies the ISO pen width of an ISO hatch pattern.

```
(apa-get-ISOPenWidth Object)
(apa-set-ISOPenWidth Object ISOPenWidth)
```

Arguments

Object	The Hatch object to which this property applies.
ISOPenWidth	An integer. One of the following:

acPenWidth000:	0.00 mm
acPenWidth013;	0.13 mm

acPenWidth018:	0.18 mm
acPenWidth025:	0.25 mm
acPenWidth035:	0.35 mm
acPenWidth050:	0.50 mm
acPenWidth070:	0.70 mm
acPenWidth100:	1.00 mm
acPenWidth140:	1.40 mm
acPenWidth200:	2.00 mm
acPenWidthUnk:	unknown

Returns

An integer.

IsPeriodic Property

Determines if the given spline is periodic (closed).

(apa-get-IsPeriodic *Object*)

Arguments

Object　　　　　　The Spline object to which this property applies.

Returns

T if it is, nil if it isn't.

IsPlanar Property

Determines if the given spline is Planar.

(apa-get-IsPlanar *Object*)

Arguments

Object　　　　　　The Spline object to which this property applies.

Returns

T if it is, nil if it isn't.

IsRational Property

Determines if the given spline is Rational.

(apa-get-IsRational *Object*)

Arguments

Object　　　　　　The Spline object to which this property applies.

Returns

T if it is, nil if it isn't.

IsXRef Property

Determines if the given block is a XRef.

(apa-get-IsXRef *Object*)

Arguments

Object　　　　　　The Block object to which this property applies.

Returns

T if the object is a XRef block, nil that it isn't.

KeyboardAccelerator Property

Specifies the Windows standard or AutoCAD classic keyboard accelerators.

```
(apa-get-KeyboardAccelerator Object)
(apa-set-KeyboardAccelerator Object Mode)
```

Arguments

Object	The PreferencesUser object to which this property applies.
Mode	An integer. One of the following:

acPreferenceClassic: Uses the AutoCAD classic accelerators.
acPreferenceCustom: Uses the Windows standard accelerators.

Returns

An integer.

KeyboardPriority Property

Controls how AutoCAD applies running object snaps the input of coordinate data.

```
(apa-get-KeyboardPriority Object)
(apa-set-KeyboardPriority Object Mode)
```

Arguments

Object	The PreferencesUser object to which this property applies.
Mode	An integer. One of the following:

acKeyboardRunningObjSnap
acKeyboardEntry
acKeyboardEntryExceptScripts

Returns

An integer.

Remarks

This property is stored in the OSNAPCOORD system variable.

Knots Property

Specifies the knot vector for a spline.

```
(apa-get-Knots Object)
(apa-set-Knots Object Knots)
```

Arguments

Object	The Spline object to which this property applies.

Returns

The knot vector of the spline.

Label Property

Specifies the content and formatting of menu items as they appear to the user.

```
(apa-get-Label Object)
(apa-set-Label Object Label)
```

Arguments

Object	The PopupMenuItem object to which this method applies.
Label	A string.

Returns

A string.

Remarks

Unlike the Caption property, this property may contain DIESEL string expressions that conditionally alter the labels each time they are displayed.

LargeButtons Property

Specifies if large buttons are to be displayed in the toolbar.

```
(apa-get-LargeButtons Object)
(apa-set-LargeButtons Object Large)
```

Arguments

Object   The Toolbar or Toolbars object to which this method applies.

Large   non-nil for large buttons, nil for small.

Returns

T if large buttons are displayed, nil if they're not.

LastHeight Property

Specifies the last text height used for a text style.

```
(apa-get-LastHeight Object)
(apa-set-LastHeight Object Height)
```

Arguments

Object   The TextStyle object to which this method applies.

Height   A positive number.

Returns

A real number.

Layer Property

Specifies the layer for an object.

```
(apa-get-Layer Object)
(apa-set-Layer Object Layer)
```

Arguments

Object   The All Drawing objects, AttributeRef, or Group object to which this method applies.

Layer   A string, specifying the name of the layer.

Returns

A string.

LayerOn Property

Specifies the state of a layer.

```
(apa-get-LayerOn Object)
(apa-set-LayerOn Object LayerOn)
```

Arguments

Object   The Layer object to which this method applies.

LayerOn   non-nil to turn the layer on, nil to turn it off.

Returns

T if it's on, nil if it's off.

Layers Property

Returns the Layers collection for the document.

```
(apa-get-Layers Object)
```

Arguments

Object   The Document object to which this property applies.

Returns

The Layers collection for the document, as an object.

Layout Property

Specifies the layout associated with the model space, paper space, or block object.

```
(apa-get-Layout Object)
(apa-set-Layout Object Layout)
```

Arguments

Object	The ModelSpace, PaperSpace ,or Block object to which this method applies.
Layout	The layout object that is associated with the model space, paper space, or block object.

Returns

An object.

LayoutCreateViewport Property

Specifies the automatic creation of a viewport for new layouts.

```
(apa-get-LayoutCreateViewport Object)
(apa-set-LayoutCreateViewport Object Mode)
```

Arguments

Object	The PreferencesDisplay object to which this method applies.
Mode	non-nil to turn the automatic creation on, nil to turn it off.

Returns

T if it's on, nil if it's off.

LayoutCrosshairColor Property

Specifies the color of the crosshairs, label text, and UCS icon for paper space layouts.

```
(apa-get-LayoutCrosshairColor Object)
(apa-set-LayoutCrosshairColor Object Color)
```

Arguments

Object	The PreferencesDisplay object to which this method applies.
Color	An integer specifying an OLE_COLOR. One of the following:

vbBlack
vbRed
vbYellow
vbGreen
vbCyan
vbBlue
vbMagenta
vbWhite

Returns

An integer.

Remarks

To specify the crosshairs color for the model space layout, use the ModelCrosshairColor property.

LayoutDisplayMargins Property

Controls the display of margins in layouts.

```
(apa-get-LayoutDisplayMargins Object)
(apa-set-LayoutDisplayMargins Object Mode)
```

Arguments

Object   The PreferencesDisplay object to which this method applies.

Mode   non-nil to turn on the display, nil to turn it off.

Returns

T if it's on, nil if it's off.

LayoutDisplayPaper Property

Controls the display of the paper background in layouts.

(apa-get-LayoutDisplayPaper Object)
(apa-set-LayoutDisplayPaper Object Mode)

Arguments

Object   The PreferencesDisplay object to which this method applies.

Mode   non-nil to turn on the display, nil to turn it off.

Returns

T if it's on, nil if it's off.

LayoutDisplayPaperShadow Property

Controls the display of the paper background shadow in layouts.

(apa-get-LayoutDisplayPaperShadow Object)
(apa-set-LayoutDisplayPaperShadow Object Mode)

Arguments

Object   The PreferencesDisplay object to which this method applies.

Mode   non-nil to turn on the display, nil to turn it off.

Returns

T if it's on, nil if it's off.

LayoutShowPlotSetup Property

Controls the display of the Plot Setup dialog when a new layout is created.

(apa-get-LayoutShowPlotSetup Object)
(apa-set-LayoutShowPlotSetup Object Mode)

Arguments

Object   The PreferencesDisplay object to which this method applies.

Mode   non-nil to turn on the display, nil to turn it off.

Returns

T if it's on, nil if it's off.

Layouts Property

Returns the Layouts collection for the document.

(apa-get-Layouts Object)

Arguments

Object   The Document object to which this property applies.

Returns

The Layouts collection for the document, as an object.

Left Property

Specifies the left edge of a toolbar.

```
(apa-get-Left Object)
(apa-set-Left Object Left)
```

Arguments

Object   The Toolbar object to which this property applies.

Left   A number specifying the left edge of the toolbar in pixels.

Returns

An integer.

Length Property

Returns the length of a line.

```
(apa-get-Length Object)
```

Arguments

Object   The Line object to which this property applies.

Returns

A real number.

LensLength Property

```
Specifies the lens length used in perspective viewing.
(apa-get-LensLength Object)
(apa-set-LensLength Object LensLength)
```

Arguments

Object   The PViewport object to which this property applies.

LensLength   A number specifying the lens length in millimeters.

Returns

A real number.

Remarks

This property is stored in the LENSLENGTH system variable.

LicenseServer Property

Provides a current list of client license servers available to the network license manager program.

```
(apa-get-LicenseServer Object)
```

Arguments

Object   The PreferencesFiles object to which this property applies.

Returns

A string.

Remarks

This property is stored in the ACADSERVER environment variable.

If ACADSERVER is not defined, the property will return an empty string (" ").

Limits Property

Specifies the drawing limits.

```
(apa-get-Limits Object 'LimMin 'LimMax)
(apa-set-Limits Object LimMin LimMax)
```

Arguments

Object	The Document object to which this method applies.
LimMin	2D point set to the lower-left limits.
LimMax	2D point set to the upper-right limits.

Returns

The list (*LimMin LimMax*).

LimMin	2D point set to the lower-left limits.
LimMax	2D point set to the upper-right limits.

Remarks

The **apa-get-Limits** function requires that *LimMin* and *LimMax* be quoted symbols.

The *LimMin* argument controls the LIMMIN system variable. The *LimMax* argument controls the LIMMAX system variable.

LinearScaleFactor Property

Specifies a scale factor for linear dimensioning measurements.

(apa-get-LinearScaleFactor *Object*)
(apa-set-LinearScaleFactor **Object ScaleFactor**)

Arguments

Object	The DimAligned, DimDiametric, DimOrdinate, DimRadial, or DimRotated object to which this property applies.
ScaleFactor	A non-zero number.

Returns

A real number.

Remarks

This property overrides the value of the DIMLFAC system variable for the given dimension.

When LinearScaleFactor is assigned a negative value, the (absolute value of the) factor is applied only in paper space.

LineSpacingFactor Property

Specifies the line spacing factor for the MText object.

(apa-get-LineSpacingFactor *Object*)
(apa-set-LineSpacingFactor *Object Factor*)

Arguments

Object	The MText object to which this property applies.
Factor	A number in the range $0.25 \leq Factor \leq 4$.

Returns

A real number.

LineSpacingStyle Property

Specifies the line spacing style for the MText object.

(apa-get-LineSpacingStyle *Object*)
(apa-set-LineSpacingStyle *Object Style*)

Arguments

Object	The MText object to which this property applies.
Style	An integer. One of the following: **acLineSpacingStyleAtLeast** **acLineSpacingStyleExactly**

Returns

An integer.

Remarks

The **acLineSpacingStyleAtLeast** setting allows the spacing between different lines of text to adjust automatically, based on the height of the largest character each line.

The **acLineSpacingStyleExactly** setting sets uniform line spacing.

Linetype Property

Specifies the linetype for an object.

```
(apa-get-Linetype Object)
(apa-set-Linetype Object Linetype)
```

Arguments

Object	The All Drawing objects, AttributeRef, Group, or Layer object to which this method applies.
Linetype	A string, specifying the name of the linetype.

Returns

A string.

LinetypeGeneration Property

Specifies how linetype patterns generate around the vertices of a 2D polyline or lightweight polyline.

```
(apa-get-LinetypeGeneration Object)
(apa-set-LinetypeGeneration Object Continuous)
```

Arguments

Object	The LightweightPolyline or Polyline object to which this method applies.
Continuous	non-nil to generate the linetype in a continuous pattern around the vertices of the polyline. nil to generate the polyline to start and end with a dash at each vertex.

Returns

T if the linetype generation is continuous, nil otherwise.

Remarks

This property overrides the value of the PLINEGEN system variable for the given polyline.

Does not apply to polylines with tapered segments.

Linetypes Property

Returns the Linetypes collection for the document.

```
(apa-get-Linetypes Object)
```

Arguments

Object	The Document object to which this property applies.

Returns

The Linetypes collection for the document, as an object.

LinetypeScale Property

Specifies the linetype scale of an Object.

```
(apa-get-LinetypeScale Object)
(apa-set-LinetypeScale Object Scale)
```

Arguments

Object	The All Drawing objects AttributeRef, or Group object to which this property applies.
Scale	A positive, non-zero number.

Returns

A real number.

Remarks

This property overrides the value of the CELTSCALE system variable for the given dimension.

LineWeight Property

Specifies the LineWeight of an individual object or the default LineWeight for the drawing.

```
(apa-get-LineWeight Object)
(apa-set-LineWeight Object LineWeight)
```

Arguments

Object   The All Drawing objects; DatabasePreferences, or Layer object to which this property applies.

LineWeight   One of the following:

```
acLnWtByLayer
acLnWtByBlock
acLnWtByLwDefault
acLnWt000
acLnWt005
acLnWt009
acLnWt013
acLnWt015
acLnWt018
acLnWt020
acLnWt025
acLnWt030
acLnWt035
acLnWt040
acLnWt050
acLnWt053
acLnWt060
acLnWt070
acLnWt080
acLnWt090
acLnWt100
acLnWt106
acLnWt120
acLnWt140
acLnWt158
acLnWt200
acLnWt211
```

Returns

An integer.

Remarks

This property overrides the value of the LWDEFAULT system variable for the given object.
LineWeight settings are visible only when the LineWeightDisplay property is set to T.

LineWeightDisplay Property

Controls the display of the LineWeights in ModelSpace.

```
(apa-get-LineWeightDisplay Object)
(apa-set-LineWeightDisplay Object Enable)
```

Arguments

Object	The DatabasePreferences object to which this property applies.
Enable	non-nil enables, nil disables.

Returns

T if enabled. nil otherwise.

LoadAcadLspInAllDocuments Property

Specifies whether *acad.lsp* is loaded only at startup or with each drawing.

```
(apa-get-LoadAcadLspInAllDocuments Object)
(apa-set-LoadAcadLspInAllDocuments Object Enable)
```

Arguments

Object	The PreferencesSystem object to which this property applies.
Enable	non-nil enables, nil disables.
	If enabled, *acad.lsp* is loaded with each drawing.
	If disabled, *acad.lsp* is loaded only with the first drawing at startup.

Returns

T if enabled. nil otherwise.

Remarks

This property is stored in the ACADLSPASDOC system variable.

LocaleID Property

Returns the locale ID of the current AutoCAD session.

```
(apa-get-LocaleID Object)
```

Arguments

Object	The Application object to which this property applies.

Returns

An integer.

Remarks

The locale ID is defined by the Windows operating system. For a list of language IDs, see the *winnt.h* file included in your Windows programming environment.

Lock Property

Locks or unlocks a layer.

```
(apa-get-Lock Object)
(apa-set-Lock Object Lock)
```

Arguments

Object	The Layer object to which this property applies.
Lock	non-nil to lock the layer, nil to unlock it.

Returns

T if locked. nil if unlocked.

LogFileOn Property

Controls logging to the log file.

```
(apa-get-LogFileOn Object)
(apa-set-LogFileOn Object LogFileOn)
```

Arguments

Object	The PreferencesOpenSave object to which this property applies.
LogFileOn	non-nil to enable logging, nil to disable it.

Returns

Returns

T if enabled. nil if not.

Remarks

This property is stored in the LOGFILEMODE system variable.

LogFilePath Property

Specifies the directory of the AutoCAD Log file.

```
(apa-get-LogFilePath Object)
(apa-set-LogFilePath Object Path)
```

Arguments

Object	The PreferencesFiles object to which this property applies.
Path	A string, specifying the directory of the AutoCAD Log File.

Returns

A string.

Remarks

This property is stored in the LOGFILEPATH system variable.

LowerLeftCorner Property

Returns the lower-left corner of the specified viewport.

```
(apa-get-LowerLeftCorner Object)
```

Arguments

Object	The Viewport object to which this property applies.

Returns

A 2D point representing the lower-left corner of the viewport.

Macro Property

Specifies the macro for the menu or toolbar item.

```
(apa-get-Macro Object)
(apa-set-Macro Object Macro)
```

Arguments

Object	The PopupMenuItem or ToolbarItem object to which this method applies.
Macro	A string containing the macro associated with the menu or toolbar item.

Returns

A string.

MainDictionary Property

Specifies the Main Dictionary file for spell checking.

```
(apa-get-MainDictionary Object)
(apa-set-MainDictionary Object FileName)
```

Arguments

Object	The PreferencesFiles object to which this property applies.

FileName	A string, specifying the main dictionary (*.dct*) file. One of the following:

enu	American English
ena	Australian English
ens	British English (ise)
enz	British English (ize)
ca	Catalan
cs	Czech
da	Danish
nl	Dutch (primary)
nls	Dutch (secondary)
fi	Finnish
fr	French (unaccented capitals)
fra	French (accented capitals)
de	German (Scharfes s)
ded	German (Dopple s)
it	Italian
no	Norwegian (Bokmal)
non	Norwegian (Nynorsk)
pt	Portuguese (Iberian)
ptb	Portuguese (Brazilian)
ru	Russian (infrequent io)
rui	Russian (frequent io)
es	Spanish (unaccented capitals)
esa	Spanish (accented capitals)
sv	Swedish

Returns

A string.

This property is stored in the DCTMAIN system variable.

The full path is not returned because this file is expected to reside in the *support* directory. The extension must be *.dct*.

MajorAxis Property

Specifies the major axis for an ellipse.

```
(apa-get-MajorAxis Object)
(apa-set-MajorAxis Object MajorAxis)
```

Arguments

Object	The Ellipse object to which this method applies.
MajorAxis	A vector specifying the major axis. The vector originates at the ellipse center.

Returns

A 3D vector.

MajorRadius Property

Specifies the length of the major axis for an ellipse.

```
(apa-get-MajorRadius Object)
(apa-set-MajorRadius Object MajorRadius)
```

Arguments

Object	The Ellipse object to which this method applies.
MajorRadius	A number specifying the length of the major axis.

Returns

A real number.

MaxActiveViewports Property

Specifies the maximum number of active viewports.

```
(apa-get-MaxActiveViewports Object)
(apa-set-MaxActiveViewports Object Max)
```

Arguments

Object   The DatabasePreferences object to which this property applies.

Max   A number in the range $2 \leq Max \leq 48$, specifying the maximum number of active viewports.

Returns

An integer.

Remarks

This property is stored in the MAXACTVP system variable, which may be set to an number in the range $2 \leq MACACTVP \leq 64$.

MaxAutoCADWindow Property

Specifies if AutoCAD should be started in a maximized window.

```
(apa-get-MaxAutoCADWindow Object)
(apa-set-MaxAutoCADWindow Object Maximize)
```

Arguments

Object   The PreferencesDisplay object to which this property applies.

Maximize   non-nil to maximize the AutoCAD window on startup, nil to not.

Returns

T if maximized on startup, nil if not.

MClose Property

Specifies if the PolygonMesh is closed in the M direction.

```
(apa-get-MClose Object)
(apa-set-MClose Object Close)
```

Arguments

Object   The PolygonMesh object to which this property applies.

Close   non-nil to close the mesh in the M direction, nil to open it.

Returns

T if closed, nil if opened.

MDensity Property

Specifies the surface density of a PolygonMesh in the M direction.

```
(apa-get-MDensity Object)
(apa-set-MDensity Object Density)
```

Arguments

Object   The PolygonMesh object to which this property applies.

Density   A number in the range $2 \leq Density \leq 256$

Returns

An integer.

Remarks

The M surface density is the number of vertices in the M direction for PolygonMesh objects that are of the following types: **acQuadSurfaceMesh**, **acCubicSurfaceMesh**, or **acBezierSurfaceMesh**.

The initial value for this property is derived from the value in the SURFU system variable + 1.
Changing the value does not seem to update the mesh.

Measurement Property

Returns the measurement for the dimension.

```
(apa-get-Measurement Object)
```

Arguments

Object   The Dim3PointAngular, DimAligned, DimAngular, DimDiametric, DimOrdinate,
                             DimRadial, or DimRotated object to which this property applies.

Returns

A real number, specifying the length for non-angular dimensions, and the angle (in radians) for angular dimensions.

MenuBar Property

Returns the MenuBar object for the application.

```
(apa-get-MenuBar Object)
```

Arguments

Object   The Application object to which this property applies.

Returns

The MenuBar object for the application.

MenuFile Property

Specifies the location of the AutoCAD menu file.

```
(apa-get-MenuFile Object)
(apa-set-MenuFile Object FileName)
```

Arguments

Object   The PreferencesFiles object to which this property applies.
FileName   A string, specifying the location of the AutoCAD menu file.

Returns

A string.

Remarks

This property contains the file name for an *.mnu*, *.mns*, or *.mnc* file. It *may* contain the drive and path, and does *not*
include the extension.

MenuFileName Property

Returns the menu file name in which the menu group is located.

```
(apa-get-MenuFileName Object)
```

Arguments

Object   The MenuGroup object to which this property applies.

Returns

A string, specifying the full path to the *.mnc* file, including extension.

MenuGroups Property

Returns the MenuGroups collection for the session.

```
(apa-get-MenuGroups Object)
```

Arguments

Object   The Application object to which this property applies.

Returns

The MenuGroups collection for the session.

Menus Property

Returns the PopupMenus collection for the menu group.

```
(apa-get-Menus Object)
```

Arguments

Object   The MenuGroup object to which this property applies.

Returns

The PopupMenus collection for the menu group.

MinorAxis Property

Returns the minor axis for an ellipse.

```
(apa-get-MinorAxis Object)
```

Arguments

Object   The Ellipse object to which this method applies.

Returns

MinorAxis A 3D vector specifying the minor axis. The vector originates at the ellipse center.

MinorRadius Property

Specifies the length of the minor axis for an ellipse.

```
(apa-get-MinorRadius Object)
(apa-set-MinorRadius Object MinorRadius)
```

Arguments

Object   The Ellipse object to which this method applies.
MinorRadius   A number specifying the length of the minor axis.

Returns

A real number.

Mode Property

Specifies the mode of the attribute definition.

```
(apa-get-Mode Object)
(apa-set-Mode Object Mode)
```

Arguments

Object   The Attribute object to which this method applies.
Mode   A number. The sum of any of the following:

acAttributeModeNormal
acAttributeModeInvisible
acAttributeModeConstant
acAttributeModeVerify
acAttributeModePreset

Returns

An integer.

Remarks

This property is overrides the value stored in the AFLAGS system variable.

ModelCrosshairColor Property

Specifies the color of the crosshairs, label text, and UCS icon for model space.

```
(apa-get-ModelCrosshairColor Object)
(apa-set-ModelCrosshairColor Object Color)
```

Arguments

Object	The PreferencesDisplay object to which this method applies.
Color	An integer specifying an OLE_COLOR. One of the following:

> **vbBlack**
> **vbRed**
> **vbYellow**
> **vbGreen**
> **vbCyan**
> **vbBlue**
> **vbMagenta**
> **vbWhite**

Returns

An integer.

Remarks

To specify the crosshairs color for the paper space model, use the LayoutCrosshairColor property.

ModelSpace Property

Returns the ModelSpace collection for the document.

```
(apa-get-ModelSpace Object)
```

Arguments

Object   The Document object to which this property applies.

Returns

The ModelSpace collection for the document, as an object.

ModelType Property

Returns the type of layout, model or paper space, to which a layout or plot configuration applies.

```
(apa-get-ModelType Object)
```

Arguments

Object   The Layout or PlotConfiguration object to which this property applies.

Returns

T if the plot configuration applies only to the Model Space tab, or the layout is the model space layout.
nil if the plot configuration applies to all layouts, or the layout is a paper space layout.

MomentOfInertia Property

Returns the moment of inertia of a solid or a region.

```
(apa-get-MomentOfInertia Object)
```

Arguments

Object   The 3DSolid or Region object to which this property applies.

Returns

A 2D or 3D list of reals.

MRUNumber Property

Specifies the number of most recently used files that appear in the File menu.

```
(apa-get-MRUNumber Object)
```

Arguments

Object   The PreferencesOpenSave object to which this property applies.

Returns

An integer. The number of most recently used files that appear in the File menu.

MSpace Property

Allows editing of the model from floating paper space viewports.

```
(apa-get-MSpace Object)
(apa-set-MSpace Object MSpace)
```

Arguments

Object   The Document object to which this property applies.

MSpace   non-nil to allows editing of the model, nil to disallow.

Returns

T if in model space, nil if in paper space.

Remarks

Before using the MSpace property, the ActiveSpace property must be set to acPaperSpace, and the display control of the PViewport object must be switched on by the Display property.

MVertexCount Property

Returns the vertex count In the M direction for a PolygonMesh.

```
(apa-get-MVertexCount Object)
```

Arguments

Object   The PolygonMesh object to which this property applies.

Returns

An integer.

Name Property

Specifies the name of the object.

```
(apa-get-Name Object)
(apa-set-Name Object Name)
```

Arguments

Object   The Application, Block, BlockRef, Dictionary, DimStyle, Document, ExternalReference, Group, Layer, Layout, Linetype, MenuGroup, MInsertBlock, ModelSpace, PaperSpace, PlotConfiguration, PopupMenu, Raster, RegisteredApplication, SelectionSet, Shape, TextStyle, Toolbar, ToolbarItem, UCS, View, Viewport, or XRecord object to which this property applies.

Name   A string.

Returns

A string.

Remarks

This property is read-only for the following objects: Application, Document, MenuGroup, ModelSpace, PaperSpace, SelectionSet, and TextStyle.

NameNoMnemonic Property

Returns the name of the popup menu without the underscore mnemonic.

`(apa-get-NameNoMnemonic Object)`

Arguments

Object   The PopupMenu object to which this property applies.

Returns

A string.

NClose Property

Specifies if the PolygonMesh is closed in the N direction

`(apa-get-NClose Object)`
`(apa-set-NClose Object Close)`

Arguments

Object   The PolygonMesh object to which this property applies.

Close   non-`nil` to close the mesh in the N direction, `nil` to open it.

Returns

`T` if closed, `nil` if opened.

NDensity Property

Specifies the surface density of a PolygonMesh in the N direction.

`(apa-get-NDensity Object)`
`(apa-set-NDensity Object Density)`

Arguments

Object   The PolygonMesh object to which this property applies.

Density   A number in the range $2 \le Density \le 256$.

Returns

An integer.

Remarks

The N surface density is the number of vertices in the N direction for PolygonMesh objects that are of the following types: **acQuadSurfaceMesh**, **acCubicSurfaceMesh**, or **acBezierSurfaceMesh**.

The initial value for this property is derived from the value in the SURFV system variable + 1.

Changing the value does not seem to update the mesh.

Normal Property

Specifies the three-dimensional normal unit vector for the entity.

`(apa-get-Normal Object)`
`(apa-set-Normal Object Normal)`

Arguments

Object   The Arc, Attribute, AttributeRef, BlockRef, Circle, Dim3PointAngular, DimAligned, DimAngular, DimDiametric, DimOrdinate, DimRadial, DimRotated , Ellipse, ExternalReference, Hatch, Leader, LightweightPolyline, Line, MInsertBlock, MText, Point, Polyline, Region, Shape, Solid, Text, Tolerance or Trace object to which this property applies.

Normal   A 3D normal vector in the WCS.

Returns

A 3D normal unit vector in the WCS, or `nil` if the object has no normal.

NumberOfControlPoints Property

Returns the number of control points of the spline.

```
(apa-get-NumberOfControlPoints Object)
```

Arguments

Object   The Spline object to which this property applies.

Returns

An integer.

NumberOfCopies Property

Specifies the number of copies to plot.

```
(apa-get-NumberOfCopies Object)
(apa-set-NumberOfCopies Object Copies)
```

Arguments

Object   The Plot object to which this property applies.

Copies   A positive, non-zero number.

Returns

An integer.

NumberOfFaces Property

Returns the number of faces for a PolyfaceMesh.

```
(apa-get-NumberOfFaces Object)
```

Arguments

Object   The PolyfaceMesh object to which this property applies.

Returns

An integer.

```
(setq vertices '((0 0 0)(2 0 0)(2 2 0)(0 2 0)
                 (0 0 1)(2 0 1)(2 2 1)(0 2 1)))
(setq faces    '((1 2 3 4)(1 2 6 5)(2 3 7 6) (3 4 8 7)(4 1 5 8)))
(setq meshObj (apa-AddPolyfaceMesh apModelSpace vertices faces))
(apa-get-NumberOfFaces meshObj)
```

NumberOfFitPoints Property

Returns the number of fit points of the spline.

```
(apa-get-NumberOfFitPoints Object)
```

Arguments

Object   The Spline object to which this property applies.

Returns

An integer.

NumberOfLoops Property

Returns the number of loops in the hatch boundary.

```
(apa-get-NumberOfLoops Object)
```

Arguments

Object   The Hatch object to which this property applies.

Returns

An integer.

NumberOfVertices Property

Returns the number of vertices for a PolyfaceMesh.

```
(apa-get-NumberOfVertices Object)
```

Arguments

Object   The PolyfaceMesh object to which this property applies.

Returns

An integer.

```
(setq vertices '((0 0 0)(2 0 0)(2 2 0)(0 2 0)
                 (0 0 1)(2 0 1)(2 2 1)(0 2 1)))
(setq faces    '((1 2 3 4)(1 2 6 5)(2 3 7 6) (3 4 8 7)(4 1 5 8)))
(setq meshObj (apa-AddPolyfaceMesh apModelSpace vertices faces))
(apa-get-NumberOfVertices meshObj)
```

NVertexCount Property

Returns the vertex count in the N direction for a PolygonMesh.

```
(apa-get-NVertexCount Object)
```

Arguments

Object   The PolygonMesh object to which this property applies.

Returns

An integer.

ObjectArxPath Property

Specifies the location for ObjectARX applications.

```
(apa-get-ObjectArxPath Object)
(apa-set-ObjectArxPath Object Path)
```

Arguments

Object   The PreferencesFiles object to which this property applies.

Path   A string, specifying the search path for ObjectARX applications.

Returns

A string.

ObjectID Property

Returns the object ID of the object.

```
(apa-get-ObjectID Object)
```

Arguments

Object   The All Drawing objects, Block, Blocks, Dictionary, Dictionaries, DimStyle, Dimstyles, Group, Groups, Layer, Layers, Layout, Layouts, Linetype, Linetypes, ModelSpace Collection, PaperSpace Collection, PlotConfiguration, PlotConfigurations, RegisteredApplication, RegisteredApplications, TextStyle, TextStyles, UCS, UCSs, View, Views, Viewport, Viewports, or XRecord object to which this property applies.

Returns

An integer.

ObjectName Property

Returns the AutoCAD class name of the object.

```
(apa-get-ObjectName Object)
```

Arguments

Object The All Drawing objects, Block, Blocks, Dictionary, Dictionaries, DimStyle, Dimstyles, Group, Groups, Layer, Layers, Layout, Layouts, Linetype, Linetypes, ModelSpace Collection, PaperSpace Collection, PlotConfiguration, PlotConfigurations, RegisteredApplication, RegisteredApplications, TextStyle, TextStyles, UCS, UCSs, View, Views, Viewport, Viewports or XRecord object to which this property applies.

Returns

A string.

ObjectSnapMode Property

Enables running Object Snaps.

```
(apa-get-ObjectSnapMode Object)
(apa-set-ObjectSnapMode Object Enable)
```

Arguments

Object The Document object to which this property applies.

Enable non-nil to enable running object snaps, nil to disable.

Returns

T if object snaps are enabled. nil otherwise.

ObjectSortByPlotting Property

Controls sorting of drawing objects by plotting order.

```
(apa-get-ObjectSortByPlotting Object)
(apa-set-ObjectSortByPlotting Object Enable)
```

Arguments

Object The DatabasePreferences object to which this property applies.

Enable non-nil to enable sorting, nil to disable.

Returns

T if enabled, nil if disabled.

Remarks

This property is stored in the SORTENTS system variable.

AutoCAD initially enables sorting for only plotting and PostScript output. Setting additional sorting options will result in slower regeneration and redrawing times.

ObjectSortByPSOutput Property

Controls sorting of drawing objects by PostScript output order.

```
(apa-get-ObjectSortByPSOutput Object)
(apa-set-ObjectSortByPSOutput Object Enable)
```

Arguments

Object The DatabasePreferences object to which this property applies.

Enable non-nil to enable sorting, nil to disable.

Returns

T if enabled, nil if disabled.

Remarks

This property is stored in the SORTENTS system variable.

AutoCAD initially enables sorting for only plotting and PostScript output. Setting additional sorting options will result in slower regeneration and redrawing times.

ObjectSortByRedraws Property

Controls sorting of drawing objects by redraw order.

```
(apa-get-ObjectSortByRedraws Object)
(apa-set-ObjectSortByRedraws Object Enable)
```

Arguments

Object   The DatabasePreferences object to which this property applies.

Enable   non-nil to enable sorting, nil to disable.

Returns

T if enabled, nil if disabled.

Remarks

This property is stored in the SORTENTS system variable.

AutoCAD initially enables sorting for only plotting and PostScript output. Setting additional sorting options will result in slower regeneration and redrawing times.

ObjectSortByRegens Property

Controls sorting of drawing objects by regen order.

```
(apa-get-ObjectSortByRegens Object)
(apa-set-ObjectSortByRegens Object Enable)
```

Arguments

Object   The DatabasePreferences object to which this property applies.

Enable   non-nil to enable sorting, nil to disable.

Returns

T if enabled, nil if disabled.

Remarks

This property is stored in the SORTENTS system variable.

AutoCAD initially enables sorting for only plotting and PostScript output. Setting additional sorting options will result in slower regeneration and redrawing times.

ObjectSortBySelection Property

Controls sorting of drawing objects by selection order.

```
(apa-get-ObjectSortBySelection Object)
(apa-set-ObjectSortBySelection Object Enable)
```

Arguments

Object   The DatabasePreferences object to which this property applies.

Enable   non-nil to enable sorting, nil to disable.

Returns

T if enabled, nil if disabled.

Remarks

This property is stored in the SORTENTS system variable.

AutoCAD initially enables sorting for only plotting and PostScript output. Setting additional sorting options will result in slower regeneration and redrawing times.

ObjectSortBySnap Property

Controls sorting of drawing objects by object snap order.

```
(apa-get-ObjectSortBySnap Object)
(apa-set-ObjectSortBySnap Object Enable)
```

Arguments

Object   The DatabasePreferences object to which this property applies.

Enable   non-nil to enable sorting, nil to disable.

Returns

T if enabled, nil if disabled.

Remarks

This property is stored in the SORTENTS system variable.

AutoCAD initially enables sorting for only plotting and PostScript output. Setting additional sorting options will result in slower regeneration and redrawing times.

ObliqueAngle Property

Specifies the oblique angle property for the object.

```
(apa-get-ObliqueAngle Object)
(apa-set-ObliqueAngle Object Angle)
```

Arguments

Object   The Attribute, AttributeRef, Shape, Text, or TextStyle object to which this property applies.

Angle   A number specifying the oblique angle in radians

Returns

A real number.

OLELaunch Property

Controls launching the parent application when plotting OLE objects.

```
(apa-get-OLELaunch Object)
(apa-set-OLELaunch Object Enable)
```

Arguments

Object   The DatabasePreferences object to which this property applies.

Enable   non-nil to enable launching, nil to disable.

Returns

T if enabled, nil if disabled.

Remarks

This property is stored in the OLESTARTUP system variable.

Plotting from the parent application achieves a higher quality plot at the cost of lower speed.

OLEQuality Property

Specifies the plot quality of OLE objects.

```
(apa-get-OLEQuality Object)
(apa-set-OLEQuality Object Quality)
```

Arguments

Object   The PreferencesOutput object to which this property applies.

Quality   One of the following:

```
acOQLineArt
acOQText
acOQGraphics
acOQPhoto
acOQHighPhoto
```

Returns

An integer.

Remarks

This property is stored in the OLEQUALITY system variable.

OnMenuBar Property

Determines if the specified popup menu is on the menu bar.

`(apa-get-OnMenuBar Object)`

Arguments

Object   The PopupMenu object to which this property applies.

Returns

T if the popup menu object is on the menu bar, nil if it isn't.

OpenSave Property

Returns the PreferencesOpenSave object.

`(apa-get-OpenSave Object)`

Arguments

Object   The Preferences object to which this property applies.

Returns

The PreferencesOpenSave object.

Remarks

There are two types of options on the Open and Save tab of the Options dialog box:

- Those that are stored in the current drawing, affecting only that drawing. These are held in the DatabasePreferences object.

- Those that are stored in the system registry, affecting the AutoCAD application. These are held in the PreferencesOpenSave object.

The following options from the Open and Save tab are stored in the DatabasePreferences object: **XrefEdit** and **XrefLayerVisibility**.

Origin Property

Specifies the origin of the UCS, block, layout, or raster image in WCS coordinates.

`(apa-get-Origin Object)`
`(apa-set-Origin Object Origin)`

Arguments

Object   The Block, Layout, PlotConfiguration, Raster, or UCS object to which this property applies.

Origin   A WCS point specifying the origin of the object.

Returns

A 2D point for Layout and PlotConfiguration objects, a 3D point for all others.

OrthoOn Property

Controls the status of Ortho mode for the viewport.

```
(apa-get-OrthoOn Object)
(apa-set-OrthoOn Object Enable)
```

Arguments

Object   The Viewport object to which this property applies.

Enable   non-nil to enable ortho mode, nil to disable.

Returns

T if enabled, nil if it not.

Remarks

In the initial release of AutoCAD 2000, applies only to active viewport.

Output Property

Returns the PreferencesOutput object.

```
(apa-get-Output Object)
```

Arguments

Object   The Preferences object to which this property applies.

Returns

The PreferencesOutput object.

Remarks

There are two types of options on the Plotting tab of the Options dialog box:

- Those that are stored in the current drawing, affecting only that drawing. These are held in the DatabasePreferences object.

- Those that are stored in the system registry, affecting the AutoCAD application. These are held in the PreferencesPlotting object.

The following option from the Plotting tab is stored in the DatabasePreferences object: **OLELaunch**.

PaperSpace Property

Returns the PaperSpace collection for the document.

```
(apa-get-PaperSpace Object)
```

Arguments

Object   The Document object to which this property applies.

Returns

The PaperSpace collection for the document, as an object.

PaperUnits Property

Specifies the units for the display of layout or plot configuration properties.

```
(apa-get-PaperUnits Object)
(apa-set-PaperUnits Object PaperUnits)
```

Arguments

Object   The Layout or PlotConfiguration object to which this property applies.

PaperUnits   One of the following:

acInches
acMillimeters
acPixels

Returns

An integer.

Remarks

This property determines the units for the display of the layout or plot configuration in the user interface.

This property does not determine the units for input or query of the ActiveX Automation properties. All ActiveX Automation properties are represented in drawing units or radians, regardless of the units settings.

Changes will not be visible until the drawing is regenerated. Use the Regen method to regenerate the drawing.

Parent Property

Returns the parent of an object.

```
(apa-get-Parent Object)
```

Arguments

Object   The MenuBar, MenuGroup,, MenuGroups, PopupMenu, PopupMenus, PopupMenuItem, Toolbar, Toolbars, or ToolbarItem object to which this property applies.

Returns

An object.

Path Property

Determines the path of the document, application, or external reference.

```
(apa-get-Path Object)
(apa-set-Path Object Path)
```

Arguments

Object   The Application, Document, or ExternalReference object to which this property applies.

Path   A string, specifying the drive and directory of the external reference. You cannot change the path to an application or document.

Returns

A string, specifying the drive and directory of the application, document, or external reference.

PatternAngle Property

Specifies the angle of the hatch pattern.

```
(apa-get-PatternAngle Object)
(apa-set-PatternAngle Object PatternAngle)
```

Arguments

Object   The Hatch object to which this property applies.

Angle   A number specifying the angle in radians.

Returns

A real number.

PatternDouble Property

Specifies if the user-defined hatch is double-hatched.

```
(apa-get-PatternDouble Object)
(apa-set-PatternDouble Object Double)
```

Arguments

Object   The Hatch object to which this property applies.

Double   non-nil to double-hatch, nil to not.

Returns

T if double-hatched, nil if not.

Remarks

If the PatternType property is set to **acHatchPatternTypePreDefined** or **acHatchPatternTypeCustomDefined**, then the PatternDouble property is not used.

PatternName Property

Returns the hatch pattern name.

```
(apa-get-PatternName Object)
```

Arguments

Object   The Hatch object to which this property applies.

Returns

A string, specifying the hatch pattern name.

PatternScale Property

Specifies the scale of the hatch pattern.

```
(apa-get-PatternScale Object)
(apa-set-PatternScale Object Scale)
```

Arguments

Object   The Hatch object to which this property applies.

Scale   A positive, non-zero number specifying the scale.

Returns

A real number.

PatternSpace Property

Specifies the spacing of for user-defined hatch patterns.

```
(apa-get-PatternSpace Object)
(apa-set-PatternSpace Object Spacing)
```

Arguments

Object   The Hatch object to which this property applies.

Spacing   A positive, non-zero number specifying the spacing.

Returns

A real number.

PatternType Property

Returns the pattern type used for the hatch.

```
(apa-get-PatternType Object)
(apa-set-PatternType Object Spacing)
```

Arguments

Object   The Hatch object to which this property applies.

Returns

An integer.   One of the following values:
                     acHatchPatternTypePredefined
                     acHatchPatternTypeUserDefined
                     acHatchPatternTypeCustomDefined

Perimeter Property

Returns the total length of the inner and outer region loops.

```
(apa-get-Perimeter Object)
```

Arguments

Object   The Region object to which this property applies.

Returns

A real number, specifying the perimeter in drawing units.

PickAdd Property

Determines if objects are added to the selection set without using the Shift key.

```
(apa-get-PickAdd Object)
(apa-set-PickAdd Object PickAdd)
```

Arguments

Object   The PreferencesSelection object to which this property applies.

PickAdd   If non-nil, adds objects to the selection set by picking. Holding down the [SHIFT]
                        key when picking removes objects from the current selection set. This is 'standard'
                        AutoCAD operation.

                        If nil, replaces the current selection set by picking. Holding down the [SHIFT] key
                        when picking adds objects from the current selection set. This is 'standard' Windows
                        operation.

Returns

T if shift-picking removes objects, nil if shift-picking adds objects.

Remarks

This property is stored in the PICKADD system variable.

PickAuto Property

Controls implied windowing.

```
(apa-get-PickAuto Object)
(apa-set-PickAuto Object PickAuto)
```

Arguments

Object   The PreferencesSelection object to which this property applies.

PickAuto   non-nil to enables implied windowing. nil to disable.

Returns

T if enabled, nil if disabled.

Remarks

This property is stored in the PICKAUTO system variable.

PickBoxSize Property

Specifies the size of the pick box.

```
(apa-get-PickBoxSize Object)
(apa-set-PickBoxSize Object Size)
```

Arguments

Object   The PreferencesDrafting object to which this property applies.

Size   An integer specifying the size in pixels.

Returns

An integer.

Remarks

This property is stored in the APERTURE system variable.

PickDrag Property

Controls the method of drawing a selection window.

```
(apa-get-PickDrag Object)
(apa-set-PickDrag Object PickDrag)
```

Arguments

Object   The PreferencesSelection object to which this property applies.

PickDrag   non-nil to specify a selection window by dragging. This is 'standard' Windows operation.

        nil to specify a selection window by two points. This is 'standard' AutoCAD operation.

Returns

T if by dragging, nil if not.

Remarks

This property is stored in the PICKDRAG system variable.

PickFirst Property

Controls noun-verb object selection.

```
(apa-get-PickFirst Object)
(apa-set-PickFirst Object PickFirst)
```

Arguments

Object   The PreferencesSelection object to which this property applies.

PickFirst   non-nil to enable noun-verb object selection. nil to disable.

Returns

T if enabled, nil if not.

Remarks

This property is stored in the PICKFIRST system variable.

PickFirstSelectionSet Property

Returns the pickfirst selection set.

```
(apa-get-PickFirstSelectionSet Object)
```

Arguments

Object   The Document object to which this property applies.

Returns

The pickfirst selection set object.

PickGroup Property

Determines if picking a single object in a group selects the entire group.

```
(apa-get-PickGroup Object)
(apa-set-PickGroup Object PickGroup)
```

Arguments

Object   The PreferencesSelection object to which this property applies.

PickGroup   non-nil to enable group picking. nil to disable.

Returns

T if enabled, `nil` if not.

Remarks

This property is stored in the PICKSTYLE system variable.

Plot Property

Returns the plot object for the document.

```
(apa-get-Plot Object)
```

Arguments

`Object`   The Document object to which this property applies.

Returns

The plot object.

PlotConfigurations Property

Returns the PlotConfigurations collection for the document.

```
(apa-get-PlotConfigurations Object)
```

Arguments

`Object`   The Document object to which this property applies.

Returns

The PlotConfigurations collection for the document, as an object.

PlotHidden Property

Controls hidden line removal when plotting

```
(apa-get-PlotHidden Object)
(apa-set-PlotHidden Object PlotHidden)
```

Arguments

`Object`   The Layout or PlotConfiguration object to which this property applies.

`PlotHidden`   non-`nil` to enable hidden line removal. `nil` to disable.

Returns

T if enabled, `nil` if not.

PlotLegacy Property

Enables legacy plot scripts.

```
(apa-get-PlotLegacy Object)
(apa-set-PlotLegacy Object PlotLegacy)
```

Arguments

`Object`   The PreferencesOutput object to which this property applies.

`PlotHidden`   non-`nil` to enable legacy plot scripts. `nil` to disable.

Returns

T if enabled, `nil` if not.

PlotOrigin Property

Specifies the plot origin.

```
(apa-get-PlotOrigin Object)
(apa-set-PlotOrigin Object PlotOrigin)
```

Arguments

Object The Layout or PlotConfiguration object to which this property applies.

PlotOrigin A 2D point specifying the plot the origin relative to the lower-left corner of the media in millimeters.

Returns

A 2D point.

Remarks

Changes will not be visible until the drawing is regenerated.

PlotPolicy Property

```
(apa-get-PlotPolicy Object)
(apa-set-PlotPolicy Object PlotPolicy)
```

Arguments

Object The PreferencesOutput object to which this property applies.

PlotPolicy One of the following:

acPolicyLegacy: An object's plot style name is associated with its color per the naming convention ACIx, where x is the color number of the object according to the AutoCAD Color Index.

acPolicyNamed: No association is made between color and plot style name. Plot style name for each object is set to the default defined in DefaultPlotStyleForObjects.

Returns

An integer.

Remarks

This property is stored in the PSTYLEPOLICY system variable.

PlotRotation Property

Specifies the rotation angle for the layout or plot configuration.

```
(apa-get-PlotRotation Object)
(apa-set-PlotRotation Object PlotRotation)
```

Arguments

Object The Layout or PlotConfiguration object to which this property applies.

PlotRotation One of the following:

ac0degrees
ac90degrees
ac180degrees
ac270degrees

Returns

An integer.

Remarks

Changes will not be visible until the drawing is regenerated.

PlotStyleName Property

Specifies the plot style name for an object, group of objects, or layer.

```
(apa-get-PlotStyleName Object)
(apa-set-PlotStyleName Object Name)
```

Arguments

Object The All Drawing objects, Group, or Layer object to which this property applies.

Name A string, specifying the plot style name for the object.

Returns

A string.

Plottable Property

Specifies if the layer is plottable.

```
(apa-get-Plottable Object)
(apa-set-Plottable Object Plottable)
```

Arguments

Object The PreferencesOutput object to which this property applies.

Plottable non-nil if the layer is plottable. nil if not.

Returns

T if plottable, nil if not.

PlotType Property

Specifies the type of plot for the layout or plot configuration.

```
(apa-get-PlotType Object)
(apa-set-PlotType Object PlotType)
```

Arguments

Object The Layout or PlotConfiguration object to which this property applies.

PlotType One of the following:

acDisplay
acExtents
acLimits
acView
acWindow
acLayout

Returns

An integer.

Remarks

Changes will not be visible until the drawing is regenerated.

PlotViewportBorders Property

Specifies if the viewport borders are to be plotted.

```
(apa-get-PlotViewportBorders Object)
(apa-set-PlotViewportBorders Object Enable)
```

Arguments

Object The Layout or PlotConfiguration object to which this property applies.

Enable non-nil to enable plotting, nil to disable.

Returns

T if enabled, nil if disabled.

PlotViewportsFirst Property

Specifies if all geometry in paper space viewports is plotted first.

```
(apa-get-PlotViewportsFirst Object)
(apa-set-PlotViewportsFirst Object Enable)
```

Arguments

Object   The Layout or PlotConfiguration object to which this property applies.

Enable   non-nil to enable plotting viewports first, nil to disable.

Returns

T if enabled, nil if disabled.

PlotWithLineWeights Property

Specifies if objects plot with the LineWeights they're assigned in the plot file.

```
(apa-get-PlotWithLineWeights Object)
(apa-set-PlotWithLineWeights Object Enable)
```

Arguments

Object   The Layout or PlotConfiguration object to which this property applies.

Enable   non-nil to specify plotting with the LineWeights in the plot style. nil to specify plotting with the LineWeights in the drawing file.

Returns

T if plotting with the LineWeights in the plot style, nil plotting with the LineWeights in the drawing file.

PlotWithPlotStyles Property

Specifies if objects plot with the configuration they're assigned in the plot file.

```
(apa-get-PlotWithPlotStyles Object)
(apa-set-PlotWithPlotStyles Object Enable)
```

Arguments

Object   The Layout or PlotConfiguration object to which this property applies.

Enable   non-nil to specify plotting with the configuration in the plot style. nil to specify plotting with the configuration in the drawing file.

Returns

T if plotting with the configuration in the plot style, nil plotting with the configuration in the drawing file.

PolarTrackingVector Property

Controls the display of polar tracking vectors.

```
(apa-get-PolarTrackingVector Object)
(apa-set-PolarTrackingVector Object Enable)
```

Arguments

Object   The PreferencesDrafting object to which this property applies.

Enable   non-nil to enable; nil to disable.

Returns

T if enabled, nil if disabled.

Remarks

This property is stored in the TRACKPATH system variable.

PostScriptPrologFile Property

Specifies a name for a customized prolog section in the *acad.psf* file.

```
(apa-get-PostScriptPrologFile Object)
(apa-set-PostScriptPrologFile Object Section)
```

Arguments

Object   The PreferencesFiles object to which this property applies.

Section   A string, specifying the section of the *acad.psf* file to use for the PSOUT command.

Returns

A string.

Remarks

This property is stored in the PSPROLOG system variable.

Preferences Property

Returns the preferences object for the specified application or document.

(apa-get-Preferences *Object*)

Arguments

Object   The Application or Document object to which this property applies.

Returns

The Preferences or DatabasePreferences object.

Preset Property

Specifies if the attribute or attribute reference is preset.

(apa-get-Preset *Object*)
(apa-set-Preset *Object Preset*)

Arguments

Object   The Attribute or AttributeReference object to which this property applies.

Preset   non-nil specifies the object is preset, nil that it isn't.

Returns

T if the object is preset, nil that it isn't.

PrimaryUnitsPrecision Property

Specifies the precision of primary units values in dimensions.

(apa-get-PrimaryUnitsPrecision *Object*)
(apa-set-PrimaryUnitsPrecision *Object Precision*)

Arguments

Object   The DimAligned, DimDiametric, DimOrdinate, DimRadial, DimRotated, or
                    Tolerance object to which this property applies.

Precision   An integer specifying the precision. One of the following:

acDimPrecisionZero:	0
acDimPrecisionOne:	0.0
acDimPrecisionTwo:	0.00
acDimPrecisionThree:	0.000
acDimPrecisionFour:	0.0000
acDimPrecisionFive:	0.00000
acDimPrecisionSix:	0.000000
acDimPrecisionSeven:	0.0000000
acDimPrecisionEight:	0.00000000

Returns

An integer.

Remarks

Overrides the DIMDEC system variable for the specified dimension.

PrincipalDirections Property

Returns the principal directions of a solid or region.

```
(apa-get-PrincipalDirections Object)
```

Arguments

Object   The 3DSolid or Region object to which this property applies.

Returns

A list of 2D X and Y directions for Regions.
A list of 3D X, Y, and Z directions for a 3DSolid.

PrincipalMoments Property

Returns the principal moments of a solid or region.

```
(apa-get-PrincipalMoments Object)
```

Arguments

Object   The 3DSolid or Region object to which this property applies.

Returns

A list of X and Y Moments for Regions.
A list of X, Y, and Z Moments for a 3DSolid.

PrinterConfigPath Property

Specifies the location of printer configuration files.

```
(apa-get-PrinterConfigPath Object)
(apa-set-PrinterConfigPath Object Path)
```

Arguments

Object   The PreferencesFiles object to which this property applies.
Path   A string, specifying the one or more directories containing the printer configuration files.

Returns

A string.

Remarks

Use a semicolon (;) to separate multiple directories

PrinterDescPath Property

Specifies the location of printer description files.

```
(apa-get-PrinterDescPath Object)
(apa-set-PrinterDescPath Object Path)
```

Arguments

Object   The PreferencesFiles object to which this property applies.
Path   A string, specifying the one or more directories containing the printer description files.

Returns

A string.

Remarks

Use a semicolon (;) to separate multiple directories.

PrinterPaperSizeAlert Property

Specifies alerting the user when a layout is configured with a paper size that differs from the default setting for the PC3 file.

```
(apa-get-PrinterPaperSizeAlert Object)
(apa-set-PrinterPaperSizeAlert Object Enable)
```

Arguments

Object	The PreferencesOutput object to which this property applies.
Enable	non-nil to enable alerts, nil to disable.

Returns

T if alerts are enabled,. nil if not.

Remarks

This property is stored in the PAPERALERT system variable.

PrinterSpoolAlert Property

Specifies whether to alert the user when the output to a device must be spooled through a system printer due to a conflict with the I/O port.

```
(apa-get-PrinterSpoolAlert Object)
(apa-set-PrinterSpoolAlert Object Alert)
```

Arguments

Object	The PreferencesOutput object to which this property applies.
Alert	One of the following:

acPrinterAlwaysAlert:	Always alert and log errors.
acPrinterAlertOnce:	Alert only the first error and log all errors.
acPrinterNeverAlertLogOnce:	Never alert, but log all errors.
acPrinterNeverAlert:	Never alert.

Returns

An integer.

Remarks

This property is stored in the PSPOOLALERT system variable.

PrinterStyleSheetPath Property

Specifies the location of printer style sheet files.

```
(apa-get-PrinterStyleSheetPath Object)
(apa-set-PrinterStyleSheetPath Object Path)
```

Arguments

Object	The PreferencesFiles object to which this property applies.
Path	A string, specifying the one or more directories containing the printer style sheet files.

Returns

A string.

Remarks

Use a semicolon (;) to separate multiple directories.

PrintFile Property

Specifies a file name to use for plot files.

```
(apa-get-PrintFile Object)
(apa-set-PrintFile Object FileName)
```

Arguments

Object	The PreferencesFiles object to which this property applies.
FileName	A string, specifying the file name to use for plot files.

Returns

A string.

Remarks

Use "." to use the default plot file name (the name of the drawing, with the file extension *.plt*).

PrintSpoolerPath Property

Specifies the location of printer spool files.

```
(apa-get-PrintSpoolerPath Object)
(apa-set-PrintSpoolerPath Object Path)
```

Arguments

Object   The PreferencesFiles object to which this property applies.

Path   A string, specifying the directory for the printer spool files.

Returns

A string.

PrintSpoolExecutable Property

Specifies the application to use for print spooling with AutoSpool.

```
(apa-get-PrintSpoolExecutable Object)
(apa-set-PrintSpoolExecutable Object FileName)
```

Arguments

Object   The PreferencesFiles object to which this property applies.

FileName   A string, specifying the name of the print spooler executable.

Returns

A string.

ProductOfInertia Property

Returns the product of inertia of a solid or a region.

```
(apa-get-ProductOfInertia Object)
```

Arguments

Object   The 3DSolid or Region object to which this property applies.

Returns

A 3D list of reals for 3DSolids, a real for Regions.

Profiles Property

Returns the profiles object for the document.

```
(apa-get-Profiles Object)
```

Arguments

Object   The Preferences object to which this property applies.

Returns

The PreferencesProfiles object.

PromptString Property

Specifies the prompt string of the attribute.

```
(apa-get-PromptString Object)
(apa-set-PromptString Object PromptString)
```

Arguments

Object	The Attribute object to which this property applies.
PromptString	A string.

Returns

A string.

ProxyImage Property

Controls the display of proxy objects

```
(apa-get-ProxyImage Object)
(apa-set-ProxyImage Object Mode)
```

Arguments

Object	The PreferencesOpenSave object to which this property applies.
Mode	One of the following:
	acProxyNotShow
	acProxyShow
	acProxyBoundingBox

Returns

An integer

Remarks

This property is saved in the PROXYSHOW system variable.

QuietErrorMode Property

Controls the quiet error mode for plot error reporting.

```
(apa-get-QuietErrorMode Object)
(apa-set-QuietErrorMode Object Quiet)
```

Arguments

Object	The Plot object to which this property applies.
Preset	non-`nil` enable, `nil` to disable.

Returns

`T` if enable, `nil` if not.

Remarks

Quiet error mode is essential plotting applications, such as batch plotting, that require uninterrupted application execution. If the quiet error mode is disabled, the normal error reporting mechanisms (alert boxes) are used.

RadiiOfGyration Property

Returns the product of inertia of a solid or a region.

```
(apa-get-RadiiOfGyration Object)
```

Arguments

Object	The 3DSolid or Region object to which this property applies.

Returns

A 3D list of reals for 3DSolids, 2D list of real for Regions.

Radius Property

Specifies the enclosed radius of arc or circle.

```
(apa-get-Radius Object)
(apa-set-Radius Object Radius)
```

Arguments

Object	The Arc or Circle object to which this property applies.
Radius	A positive number, specifying the radius of the object.

Returns

A real number.

RadiusRatio Property

Specifies the minor to major axis ratio for an ellipse.

```
(apa-get-Radius Object)
(apa-set-Radius Object RadiusRatio)
```

Arguments

Object	The Ellipse object to which this property applies.
RadiusRatio	A number in the range $0 < RadiusRatio \leq 1$.

Returns

A real number.

ReadOnly Property

Determines if the document is read-only.

```
(apa-get-ReadOnly Object)
```

Arguments

Object	The Document object to which this property applies.

Returns

T if the drawing is read-only, nil if it isn't.

RegisteredApplications Property

Returns the RegisteredApplications collection for the document.

```
(apa-get-RegisteredApplications Object)
```

Arguments

Object	The Document object to which this property applies.

Returns

The RegisteredApplications collection for the document, as an object.

RemoveHiddenLines Property

Controls hidden line removal when plotting a paper space viewport

```
(apa-get-RemoveHiddenLines Object)
(apa-set-RemoveHiddenLines Object Enable)
```

Arguments

Object	The PViewport object to which this property applies.
Enable	non-nil to enable hidden line removal. nil to disable.

Returns

T if enabled, nil if not.

RenderSmoothness Property

Specifies the smoothness of shaded, rendered, and hidden line-removed objects.

```
(apa-get-RenderSmoothness Object)
(apa-set-RenderSmoothness Object Smoothness)
```

Arguments

Object   The DatabasePreferences object to which this property applies.

Smoothness   A number in the range $0.01 \leq Smoothness \leq 10$.

Returns

A real number.

Remarks

This property is stored in the FACETRES system variable.

Rotation Property

Specifies the rotation angle for the object.

```
(apa-get-Rotation Object)
(apa-set-Rotation Object Rotation)
```

Arguments

Object   The Attribute, AttributeRef, BlockRef, Dim3PointAngular, DimAligned, DimAngular, DimDiametric, DimOrdinate, DimRadial, DimRotated, ExternalReference, MInsertBlock, MText, Raster, Shape, or Text object to which this property applies.

Rotation   The rotation angle in radians.

Returns

A real number.

RoundDistance Property

Specifies the rounding of units for dimensions.

```
(apa-get-RoundDistance Object)
(apa-set-RoundDistance Object RoundDistance)
```

Arguments

Object   The DimAligned, DimDiametric, DimOrdinate, DimRadial, or DimRotated object to which this property applies.

RoundDistance   A positive number specifying the value to which to round distances.

Returns

A real number.

Remarks

Overrides the DIMRND system variable for the specified dimension.

Rows Property

Specifies the number of rows in a block array.

```
(apa-get-Rows Object)
(apa-set-Rows Object RowSpacing)
```

Arguments

Object   The MInsertBlock object to which this property applies.

Rows   Positive number specifying the number of rows in the block array.

Returns

An integer.

RowSpacing Property

Specifies the spacing of rows in a block array.

```
(apa-get-Rows Object)
(apa-set-Rows Object Spacing)
```

Arguments

Object	The MInsertBlock object to which this property applies.
Spacing	Number specifying the spacing of rows in the block array.

Returns

An real number.

SaveAsType Property

Specifies the type of file to save the drawing as.

```
(apa-get-SaveAsType Object)
(apa-set-SaveAsType Object Type)
```

Arguments

Object	The PreferencesOpenSave object to which this property applies.	
Type	A number.	One of the following:
	acR12_DXF:	AutoCAD Release12/LT2 DXF (*.dxf)
	acR13_DWG:	AutoCAD Release13/LT95 DWG (*.dwg)
	acR13_DXF:	AutoCAD Release13/LT95 DXF (*.dxf)
	acR14_DWG:	AutoCAD Release14/LT97 DWG (*.dwg)
	acR14_DXF:	AutoCAD Release14/LT97 DXF (*.dxf)
	acR15_DWG:	AutoCAD 2000 DWG (*.dwg)
	acR15_DXF:	AutoCAD 2000 DXF (*.dxf)
	acR15_Template:	AutoCAD 2000 Template File (*.dwt)
	acNative:	The current drawing release format.
	acUnknown:	The drawing type is unknown.

Returns

An integer.

Saved Property

Specifies the drawing has no unsaved changes.

```
(apa-get-Saved Object)
```

Arguments

Object	The Document object to which this property applies.

Returns

T if the drawing has no unsaved changes, nil if it does.

SavePreviewThumbnail Property

Specifies if preview images are saved with the drawing.

```
(apa-get-SavePreviewThumbnail Object)
(apa-set-SavePreviewThumbnail Object Save)
```

Arguments

Object	The PreferencesOpenSave object to which this property applies.
Save	non-nil to save preview images, nil to not save.

Returns

T if preview images are to be saved, nil not.

Remarks

This property is saved in the RASTERPREVIEW system variable.

ScaleFactor Property

Specifies the scale factor for the object.

```
(apa-get-ScaleFactor Object)
(apa-set-ScaleFactor Object ScaleFactor)
```

Arguments

`Object`	The Attribute, AttributeRef, Dim3PointAngular, DimAligned, DimAngular, DimDiametric, DimOrdinate, DimRadial, DimRotated, Leader, Raster, Shape, Text, or Tolerance object to which this property applies.
`ScaleFactor`	A positive, non-zero number specifying the scale factor for the object.

Returns

A real number.

Remarks

For Dimension, Leader, and Tolerance objects, this property overrides the value in the DIMSCALE system variable. For all other objects, this is the width factor for the text.

ScaleLineWeights Property

Specifies if LineWeights are to be scaled along with the geometry when printed from paper space.

```
(apa-get-ScaleLineWeights Object)
(apa-set-ScaleLineWeights Object Scale)
```

Arguments

`Object`	The Layout or PlotConfiguration object to which this property applies.
`ScaleFactor`	A positive, non-zero number specifying the scale factor for the object.
`Scale`	non-nil to scale , nil to not scale.

Returns

T if scaled, nil if not.

Remarks

This property is stored in the LWSCALE system variable.

The property disabled for the model space layout.

SCMCommandMode Property

Determines the action when a user right-clicks in the drawing area while a command is active.

```
(apa-get-SCMCommandMode Object)
(apa-set-SCMCommandMode Object Mode)
```

Arguments

`Object`	The PreferencesUser object to which this property applies.
`Mode`	A number specifying the right-click action. One of the following:

`acEnter`	Disables the Command Short Cut Menu
`acEnableSCM`	Enables the Command Short Cut Menu
`acEnableSCMOptions`	Enables the Command SCM when options are present.

Returns

An integer.

Remarks

This property is stored in the SHORTCUTMENU system variable.

SCMDefaultMode Property

Determines the action when a user right-clicks in the drawing area at the command: prompt, with no objects are selected.

```
(apa-get-SCMDefaultMode Object)
(apa-set-SCMDefaultMode Object Mode)
```

Arguments

Object   The PreferencesUser object to which this property applies.

Mode   A number specifying the right-click action. One of the following:

acRepeatLastCommand	Disables the Default Short Cut Menu
acSCM	Enables the Default Short Cut Menu

Returns

An integer.

Remarks

This property is stored in the SHORTCUTMENU system variable.

SCMEditMode Property

Determines the action when a user right-clicks in the drawing area while at the command prompt with objects selected.

```
(apa-get-SCMEditMode Object)
(apa-set-SCMEditMode Object Mode)
```

Arguments

Object   The PreferencesUser object to which this property applies.

Mode   A number specifying the right-click action. One of the following:

acEdRepeatLastCommand	Disables the Edit Short Cut Menu
acEdSCM	Enables the Edit Short Cut Menu

Returns

An integer.

Remarks

This property is stored in the SHORTCUTMENU system variable.

SecondPoint Property

Specifies the second point through which the ray or xline passes.

```
(apa-get-SecondPoint Object)
(apa-set-SecondPoint Object SecondPoint)
```

Arguments

Object   The Ray or XLine object to which this property applies.

SecondPoint   The point through which the ray or xline passes.

Returns

A 3D point.

SegmentPerPolyline Property

Specifies the number of line segments to be generated when spline-fitting a polyline.

```
(apa-get-SegmentPerPolyline Object)
(apa-set-SegmentPerPolyline Object Segments)
```

Arguments

Object   The DatabasePreferences object to which this property applies.

Segments   A number in the range $0 \leq$ Segments ≤ 32767.

Returns

An integer

Remarks

This property is stored in the SPLINESEGS system variable.

Although allowable values for SPLINESEGS are the range of numbers $-32767 \leq$ SPLINESEGS ≤ 32767, you are limited to positive numbers while setting this property.

Selection Property

Returns the PreferencesSelection object.

(apa-get-Selection Object)

Arguments

Object   The Preferences object to which this property applies.

Returns

The PreferencesSelection object.

Remarks

All of the options on the Selection tab of the Options dialog box, are stored in the system registry, affecting the AutoCAD application. These are held in the PreferencesSelection object.

SelectionSets Property

Returns the SelectionSets collection for the document.

(apa-get-SelectionSets Object)

Arguments

Object   The Document object to which this property applies.

Returns

The SelectionSets collection for the document, as an object.

ShortcutMenu Property

Determines if the specified popup menu is the object snap cursor menu.

(apa-get-ShortcutMenu Object)

Arguments

Object   The Popup object to be queried.

Returns

T if the popup menu is the object snap cursor menu, nil if it's not.

ShortCutMenuDisplay Property

Controls the display of the ShortCutMenus when the user right-clicks in the drawing area.

(apa-get-ShortCutMenuDisplay Object)
(apa-set-ShortCutMenuDisplay Object Enable)

Arguments

Object   The PreferencesUser object to which this property applies.

Enable   non-nil enables short cut menus, nil disables.

Returns

T if enabled. nil otherwise.

Remarks

This property is stored in the SHORTCUTMENU system variable.

ShowPlotStyles Property

Determines if plot styles are to be used when plotting.

```
(apa-get-ShowPlotStyles Object)
(apa-set-ShowPlotStyles Object Enable)
```

Arguments

Object   The Layout or PlotConfiguration object to which this property applies.

Enable   non-nil enables plot styles, nil disables.

Returns

T if enabled. nil otherwise.

ShowProxyDialogBox Property

Controls the display of the proxy dialog box when you open a drawing that contains custom objects.

```
(apa-get-ShowProxyDialogBox Object)
(apa-set-ShowProxyDialogBox Object Enable)
```

Arguments

Object   The PreferencesOpenSave object to which this property applies.

Enable   non-nil enables, nil disables.

Returns

T if enabled. nil otherwise.

ShowRasterImage Property

Controls the display of raster images during real time pans and zooms.

```
(apa-get-ShowRasterImage Object)
(apa-set-ShowRasterImage Object Enable)
```

Arguments

Object   The PreferencesDisplay object to which this property applies.

Enable   non-nil enables, nil disables.

Returns

T if enabled. nil otherwise.

ShowRotation Property

Determines if raster image rotation is enabled.

```
(apa-get-ShowRotation Object)
(apa-set-ShowRotation Object Enable)
```

Arguments

Object   The Raster object to which this property applies.

Enable   non-nil enables, nil disables.

Returns

T if enabled. nil otherwise.

Remarks

This property seems to have no effect.

ShowWarningMessages Property

Resets the dialog boxes which checkmarks by the "Don't Display This Warning Again" toggle.

```
(apa-get-ShowWarningMessages Object)
(apa-set-ShowWarningMessages Object Enable)
```

Arguments

Object	The PreferencesSystem object to which this property applies.
Enable	non-nil resets, nil doesn't reset.

Returns

T if reset. nil otherwise.

SingleDocumentMode Property

Determines if AutoCAD works in SDI or MDI mode.

```
(apa-get-SingleDocumentMode Object)
(apa-set-SingleDocumentMode Object SDI)
```

Arguments

Object	The PreferencesSystem object to which this property applies.
SDI	non-nil sets SDI mode, nil sets MDI mode.

Returns

T if SDI mode. nil if MDI mode.

Remarks

If more than one drawing is currently open, this property is forced to nil.

SnapBasePoint Property

Specifies the snap base point for the viewport.

```
(apa-get-SnapBasePoint Object)
(apa-set-SnapBasePoint Object BasePoint)
```

Arguments

Object	The PViewport or Viewport object to which this property applies.
SDI	A WCS point.

Returns

A 2D WCS point.

Remarks

This property is stored in the SNAPBASE system variable.

SnapOn Property

Specifies the status of the snap grid.

```
(apa-get-SnapOn Object)
(apa-set-SnapOn Object Enable)
```

Arguments

Object	The PViewport or Viewport object to which this property applies.
Enable	non-nil enables, nil disables.

Returns

T if enabled. nil if disabled.

Remarks

This property is stored in the SNAPMODE system variable.

Once you've changed the SnapOn property of a viewport, you must reset the active viewport to see the changes.

SnapRotationAngle Property

Specifies the snap rotation angle for the viewport, relative to the current UCS.

```
(apa-get-SnapRotationAngle Object)
(apa-set-SnapRotationAngle Object Rotation)
```

Arguments

| Object | The PViewport or Viewport object to which this property applies. |
| Rotation | An angle in radians. |

Returns

A real number.

Remarks

This property is stored in the SNAPANG system variable.

SolidFill Property

Controls the display and plotting of solid fill and all hatch patterns.

```
(apa-get-SolidFill Object)
(apa-set-SolidFill Object Enable)
```

Arguments

| Object | The DatabasePreferences object to which this property applies. |
| Enable | non-nil enables solid fill and all hatch patterns., nil disables. |

Returns

T if enabled. nil if disabled.

Remarks

This property is stored in the FILLMODE system variable.

StandardScale Property

Specifies the standard scale factor for the viewport.

```
(apa-get-StandardScale Object)
(apa-set-StandardScale Object StandardScale)
```

Arguments

| Object | The Layout, PViewport, or PlotConfiguration object to which this property applies. |
| Scale | For PViewports, one of the following: |

acVpScaleToFit	Scale to fit
acVpCustomScale	Custom
acVp1_128in_1ft	1/128"= 1'
acVp1_64in_1ft	1/64"= 1'
acVp1_32in_1ft	1/32"= 1'
acVp1_16in_1ft	1/16"= 1'
acVp3_32in_1ft	3/32"= 1'
acVp1_8in_1ft	1/8" = 1'
acVp3_16in_1f	3/16"= 1'

acVp1_4in_1ft	1/4" = 1'
acVp3_8in_1ft	3/8" = 1'
acVp1_2in_1ft	1/2" = 1'
acVp3_4in_1ft	3/4" = 1'
acVp1in_1ft	1"= 1'
acVp3in_1ft	3"= 1'
acVp6in_1ft	6"= 1'
acVp1ft_1ft	1'= 1'
acVp1_1	1:1
acVp1_2	1:2
acVp1_4	1:4
acVp1_8	1:8
acVp1_10	1:10
acVp1_16	1:16
acVp1_20	1:20
acVp1_30	1:30
acVp1_40	1:40
acVp1_50	1:50
acVp1_100	1:100
acVp2_1	2:1
acVp4_1	4:1
acVp8_1	8:1
acVp10_1	10:1
acVp100_1	100:1

For Layouts and PlotConfigurations, one of the following:

acScaleToFit	Scale to Fit
ac1_128in_1ft	1/128"= 1'
ac1_64in_1ft	1/64"= 1'
ac1_32in_1ft	1/32"= 1'
ac1_16in_1ft	1/16"= 1'
ac3_32in_1ft	3/32"= 1'
ac1_8in_1ft	1/8" = 1'
ac3_16in_1ft	3/16"= 1'
ac1_4in_1ft	1/4" = 1'
ac3_8in_1ft	3/8" = 1'
ac1_2in_1ft	1/2" = 1'
ac3_4in_1ft	3/4" = 1'
ac1in_1ft	1"= 1'
ac3in_1ft	3"= 1'
ac6in_1ft	6"= 1'
ac1ft_1ft	1'= 1'
ac1_1	1:1
ac1_2	1:2
ac1_4	1:4
ac1_8	1:8
ac1_10	1:10
ac1_16	1:16
ac1_20	1:20
ac1_30	1:30
ac1_40	1:40
ac1_50	1:50
ac1_100	1:100

ac2_1	2:1
ac4_1	4:1
ac8_1	8:1
ac10_1	10:1
ac100_1	100:1

Returns

An integer.

Remarks

Changes will not be visible until the drawing is regenerated.

StartAngle Property

Specifies the Start Angle of an arc, or ellipse.

```
(apa-get-StartAngle Object)
(apa-set-StartAngle Object StartAngle)
```

Arguments

Object   The Arc or Ellipse object to which this method applies.

StartAngle   A number specifying the Start Angle in radians. Use 2p to specify a closed ellipse.

Returns

A real number.

StartParameter Property

Specifies the start parameter for an ellipse.

```
(apa-get-StartParameter Object)
(apa-set-SLartParameter Object StartParameter)
```

Arguments

Object   The Ellipse object to which this method applies.

StartParameter A number specifying the Start Parameter.

Returns

A real number.

Remarks

The start and start parameters of the ellipse are supposedly computed with on the following equation:

$$P(\theta) = A \times \cos(\theta) + B \times \sin(\theta)$$

where A and B are the semi-major and semi-minor axes respectively, and θ is the starting or starting angle of the ellipse.

In fact, this property, as set and returned by AutoCAD, seems to have the same value as the StartAngle property.

StartPoint Property

Specifies the startpoint for an arc, line, or ellipse.

```
(apa-get-StartPoint Object)
(apa-set-StartPoint Object StartPoint)
```

Arguments

Object   The Arc, Line, or Ellipse object to which this method applies.

StartPoint   A point. You may not set the startpoint for an ellipse or an arc.

Returns

A 3D point.

StartTangent Property

Specifies the start tangent of the spline as a directional vector.

```
(apa-get-StartTangent Object)
(apa-set-StartTangent Object StartTangent)
```

Arguments

Object	The Spline object to which this method applies.
StartTangent	Vector specifying the tangency of the curve at the start of the spline.

Returns

A 3D vector.

StatusID Property

Determines if the specified viewport is the active viewport.

```
(apa-get-StatusID Application Viewport)
```

Arguments

Application	The Application object to which this method applies.
Viewport	The Viewport object to which this method applies.

Returns

T if the viewport is the active viewport, nil if it isn't.

Remarks

This function seems to always return nil.

StoreSQLIndex Property

Determines if AutoCAD stores the SQL index with the drawing.

```
(apa-get-StoreSQLIndex Object)
(apa-set-StoreSQLIndex Object Store)
```

Arguments

Object	The PreferencesSystem object to which this property applies.
Store	non-nil to store, nil to not store.

Returns

T if stored. nil if not.

StyleName Property

Specifies the name of the text style for the object.

```
(apa-get-StyleName Object)
(apa-set-StyleName Object StyleName)
```

Arguments

Object	The Attribute, AttributeRef, DimAligned, Dim3PointAngular, DimAngular, DimDiametric, DimOrdinate, DimRadial, DimRotated, Leader, MLine, MText, Text, or Tolerance object to which this property applies.
StyleName	A string.

Returns

A string.

Remarks

This property overrides the value from the DIMSTYLE system variable for dimensions, from the TEXTSTYLE system variable for text, and from the CMLSTYLE variable for MLines.

StyleSheet Property

Specifies the name of the style sheet for the object.

```
(apa-get-StyleSheet Object)
(apa-set-StyleSheet Object StyleSheet)
```

Arguments

Object	The Layout or PlotConfiguration to which this property applies.
StyleSheet	A string.

Returns

A string.

SubMenu Property

Returns the popup menu associated with a submenu.

```
(apa-get-SubMenu Object)
```

Arguments

Object   The PopupMenuItem object to which this property applies.

Returns

The submenu associated with the popup menu item, or nil if the popup menu item is not a submenu.

SupportPath Property

Specifies the directories AutoCAD searches for support files.

```
(apa-get-SupportPath Object)
(apa-set-SupportPath Object Path)
```

Arguments

Object	The PreferencesFiles object to which this property applies.
Path	A string, specifying the directories in which to search.

Returns

A string.

Remarks

Semicolons separate multiple directories.

SuppressLeadingZeros Property

Specifies the suppression of leading zeros in dimension values.

```
(apa-get-SuppressLeadingZeros Object)
(apa-set-SuppressLeadingZeros Object Suppress)
```

Arguments

Object	The DimAligned, DimDiametric, DimOrdinate, DimRadial, or DimRotated object to which this property applies.
Suppress	non-nil to suppress leading zeros. nil to not-suppress.

Returns

T if leading zeros are suppressed. nil otherwise.

Remarks

Overrides the DIMZIN system variable for the specified dimension.

SuppressTrailingZeros Property

Specifies the suppression of trailing zeros in dimension values.

```
(apa-get-SuppressTrailingZeros Object)
(apa-set-SuppressTrailingZeros Object Suppress)
```

Arguments

Object	The DimAligned, DimDiametric, DimOrdinate, DimRadial, or DimRotated object to which this property applies.
Suppress	non-nil to suppress trailing zeros. nil to not-suppress.

Returns

T if trailing zeros are suppressed. nil otherwise.

Remarks

Overrides the DIMZIN system variable for the specified dimension.

SuppressZeroFeet Property

Specifies the suppression of a zero foot measurement in dimension values.

```
(apa-get-SuppressZeroFeet Object)
(apa-set-SuppressZeroFeet Object Suppress)
```

Arguments

Object	The DimAligned, DimDiametric, DimOrdinate, DimRadial, or DimRotated object to which this property applies.
Suppress	non-nil to suppress zero-feet. nil to not-suppress.

Returns

T if zero feet are suppressed. nil otherwise.

Remarks

Overrides the DIMZIN system variable for the specified dimension.

SuppressZeroInches Property

Specifies the suppression of a zero inch measurement in dimension values.

```
(apa-get-SuppressZeroInches Object)
(apa-set-SuppressZeroInches Object Suppress)
```

Arguments

Object	The DimAligned, DimDiametric, DimOrdinate, DimRadial, or DimRotated object to which this property applies.
Suppress	non-nil to suppress zero inches, nil to not-suppress.

Returns

T if zero inches are suppressed. nil otherwise.

Remarks

Overrides the DIMZIN system variable for the specified dimension.

System Property

Returns the PreferencesSystem object.

```
(apa-get-System Object)
```

Arguments

Object	The Preferences object to which this property applies.

Returns

The PreferencesSystem object.

Remarks

There are two types of options on the System tab of the Options dialog box:

- Those that are stored in the current drawing, affecting only that drawing. These are held in the DatabasePreferences object.
- Those that are stored in the system registry, affecting the AutoCAD application. These are held in the PreferencesSystem object.

The following option on the System tab is held in the DatabasePreferences object: **AllowLongSymbolNames**.

TablesReadOnly Property

Specifies whether database tables are to be opened in read-only mode.

```
(apa-get-TablesReadOnly Object)
(apa-set-TablesReadOnly Object ReadOnly)
```

Arguments

Object   The PreferencesSystem object to which this property applies.

ReadOnly   non-nil to open in read-only mode, nil to open in read-write mode.

Returns

T for read-only. nil for read-write.

TabOrder Property

Specifies the tab order for a layout.

```
(apa-get-TabOrder Object)
(apa-set-TabOrder Object TabOrder)
```

Arguments

Object   The Layout object to which this property applies.

TabOrder   The tab order number for the layout. The first tab after the model space tab is 1.

Returns

An integer.

Remarks

You cannot change the tab order (0) of the model space layout.

TagString Property

Specifies the tag string of the attribute

```
(apa-get-TagString Object)
(apa-set-TagString Object TagString)
```

Arguments

Object   The Attribute, AttributeRef, PopupMenu, PopupMenuItem, Toolbar, or ToolbarItem
                     object to which this property applies.

TagString   The tag for the Attribute or AttributeRef object.
                     You cannot set the tag for PopupMenu and Toolbar objects.

Returns

A string.

Target Property

Specifies the target of the PViewport, View, or Viewport.

```
(apa-get-Target Object)
(apa-set-Target Object Target)
```

Arguments

`Object`	The PViewport, View, or Viewport object to which this method applies.
`Target`	Point specifying the target of the object.

Returns

A point.

TempFileExtension Property

Specifies the temporary file extension.

```
(apa-get-TempFileExtension Object)
(apa-set-TempFileExtension Object FileExtension)
```

Arguments

`Object`	The PreferencesOpenSave object to which this property applies.
`FileExtension`	A string, specifying the temporary file extension.

Returns

A string.

TempFilePath Property

Specifies the directory AutoCAD uses for temporary files.

```
(apa-get-TempFilePath Object)
(apa-set-TempFilePath Object Path)
```

Arguments

`Object`	The PreferencesFiles object to which this property applies.
`Path`	A string, specifying the directory for temporary files.

Returns

A string.

TemplateDwgPath Property

Specifies the directory AutoCAD uses for temporary files.

```
(apa-get-TemplateDwgPath Object)
(apa-set-TemplateDwgPath Object Path)
```

Arguments

`Object`	The PreferencesFiles object to which this property applies.
`Path`	A string, specifying the directory for template files.

Returns

A string.

TempXRefPath Property

Specifies the directory AutoCAD uses for temporary xref files.

```
(apa-get-TempXRefPath Object)
(apa-set-TempXRefPath Object Path)
```

Arguments

Object	The PreferencesFiles object to which this property applies.
Path	A string, specifying the directory for temporary xref files.

Returns

A string.

TextAlignmentPoint Property

Specifies the text alignment point for the object.

```
(apa-get-TextAlignmentPoint Object)
(apa-set-TextAlignmentPoint Object Point)
```

Arguments

Object	The Attribute, AttributeRef, or Text object to which this property applies.
Point	A WCS point.

Returns

A 3D point, or `nil` if you attempt to set the TextAlignmentPoint property for left-justified text.

Remarks

You cannot specify a Text Alignment Point for left-justified text.

TextColor Property

Specifies the text color for dimensions and tolerances.

```
(apa-get-TextColor Object)
(apa-set-TextColor Object Color)
```

Arguments

Object	The Dim3PointAngular, DimAligned, DimAngular, DimDiametric, DimOrdinate, DimRadial, DimRotated, or Tolerance object to which this property applies.
Color	An ACI integer in the range $0 \leq Color \leq 256$, or one of the following:

acByBlock
acByLayer
acRed
acYellow
acGreen
acCyan
acBlue
acMagenta
acWhite

Returns

An integer.

Remarks

Overrides the DIMCLRT system variable.

TextEditor Property

Specifies the text editor for the MTEXT command.

```
(apa-get-TextEditor Object)
(apa-set-TextEditor Object TextEditor)
```

Arguments

Object	The PreferencesFiles object to which this property applies.
TextEditor	A string, specifying the name of the executable to be used for editing MText, or "Internal" to use the internal editor.

Returns

A string.

Remarks

This property is stored in the MTEXTED system variable.

TextFont Property

Specifies the text font in the AutoCAD command line and text window.

```
(apa-get-TextFont Object)
(apa-set-TextFont Object Font)
```

Arguments

Object	The PreferencesDisplay object to which this property applies.
Font	A string, specifying the name of the font.

Returns

A string.

TextFontSize Property

Specifies the size of the text in the AutoCAD command line and text window.

```
(apa-get-TextFontSize Object)
(apa-set-TextFontSize Object Size)
```

Arguments

Object	The PreferencesDisplay object to which this property applies.
Size	A number specifying the height of the text in points.

Returns

An Integer

TextFontStyle Property

Specifies the style of the text in the AutoCAD command line and text window.

```
(apa-get-TextFontStyle Object)
(apa-set-TextFontStyle Object Style)
```

Arguments

Object	The PreferencesDisplay object to which this property applies.
Style	A number. One of the following:

> **acFontRegular**
> **acFontItalic**
> **acFontBold**
> **acFontBoldItalic**

Returns

An Integer.

TextFrameDisplay Property

Specifies if text is to be displayed as a frame or as text.

```
(apa-get-TextFrameDisplay Object)
(apa-set-TextFrameDisplay Object Frame)
```

Arguments

Object	The DatabasePreferences object to which this property applies.
Frame	non-nil to display text frames, nil to display text.

Returns

T if displaying frames, nil if displaying text.

Remarks

This property is stored in the QTEXTMODE system variable.

You must regenerate the drawing for the changes to be visible.

TextGap Property

Specifies the gap between the dimension line and the dimension text.

```
(apa-get-TextGap Object)
(apa-set-TextGap Object TextGap)
```

Arguments

Object	The Dim3PointAngular, DimAligned, DimAngular, DimDiametric, DimOrdinate, DimRadial, DimRotated, or Leader object to which this property applies.
TextGap	The absolute value specifies the text gap. A negative value specifies Basic dimensions (box around the text).

Returns

A real number.

Remarks

Overrides the DIMGAP system variable.

TextGenerationFlag Property

Specifies the text generation for the specified object.

```
(apa-get-TextGenerationFlag Object)
(apa-set-TextGenerationFlag Object Flag)
```

Arguments

Object	The Attribute, AttributeRef, Text, or TextStyle object to which this property applies.
Flag	A number specifying the text generation. One of the following:

acTextFlagBackward
acTextFlagUpsideDown
acTextFlagBackward + acTextFlagUpsideDown

Returns

An integer.

Remarks

You cannot set the TextGenerationFlag to 'none of the above.' This is a bug.

TextHeight Property

Specifies the text height for dimensions and tolerances.

```
(apa-get-TextHeight Object)
(apa-set-TextHeight Object Height)
```

Arguments

Object	The Dim3PointAngular, DimAligned, DimAngular, DimDiametric, DimOrdinate, DimRadial, DimRotated, or Tolerance object to which this property applies.
Height	A non-zero, positive number.

Returns

A real number.

Remarks

Overrides the DIMTXT system variable.

TextInside Property

Specifies if text should be forced inside the extension lines.

```
(apa-get-TextInside Object)
(apa-set-TextInside Object Inside)
```

Arguments

Object	The Dim3PointAngular, DimAligned, DimAngular, DimDiametric, DimRadial, or DimRotated object to which this property applies.
Inside	non-nil to force text inside, nil to not.

Returns

T if text is forced inside, nil if it's not.

Remarks

Overrides the DIMTIX system variable.

TextInsideAlign Property

Specifies if text, when drawn inside the extension lines, should be aligned horizontally.

```
(apa-get-TextInsideAlign Object)
(apa-set-TextInsideAlign Object Align)
```

Arguments

Object	The Dim3PointAngular, DimAligned, DimAngular, DimDiametric, DimRadial, or DimRotated object to which this property applies.
Align	non-nil for horizontal, nil for not.

Returns

T if text is horizontal, nil if it's not.

Remarks

Overrides the DIMTIH system variable.

TextMovement Property

Specifies the dimension text movement rules.

```
(apa-get-TextMovement Object)
(apa-set-TextMovement Object TextMovement)
```

Arguments

Object	The Dim3PointAngular, DimAligned, DimAngular, DimDiametric, DimOrdinate DimRadial, or DimRotated object to which this property applies.

Movement	A number. One of the following:
	`acDimLineWithText`
	`acMoveTextAddLeader`
	`acMoveTextNoLeader`.

Returns

An integer.

Remarks

Overrides the DIMTMOVE system variable.

TextOutsideAlign Property

Specifies if text, when drawn outside the extension lines, should be aligned horizontally.

```
(apa-get-TextOutsideAlign Object)
(apa-set-TextOutsideAlign Object Align)
```

Arguments

Object	The Dim3PointAngular, DimAligned, DimAngular, DimDiametric, DimRadial, or DimRotated object to which this property applies.
Align	non-nil for horizontal, nil for not.

Returns

T if text is horizontal, nil if it's not.

Remarks

Overrides the DIMTOH system variable.

TextOverride Property

Specifies the dimension text override.

```
(apa-get-TextOverride Object)
(apa-set-TextOverride Object TextOverride)
```

Arguments

Object	The Dim3PointAngular, DimAligned, DimAngular, DimDiametric, DimOrdinate DimRadial, or DimRotated object to which this property applies.
TextOverride	A string, specifying the dimension text override.

Returns

An string.

Remarks

An empty string (" ") displays the calculated dimension string.

A pair of angle brackets (<>) represents the calculated dimension.

A pair of square brackets ([ ]) represents the calculated secondary value. Alternate units need not be enabled for this to work.

TextPosition Property

Specifies the dimension text position.

```
(apa-get-TextPosition Object)
(apa-set-TextPosition Object TextPosition)
```

Arguments

Object	The Dim3PointAngular, DimAligned, DimAngular, DimDiametric, DimOrdinate DimRadial, or DimRotated object to which this property applies.
TextPosition	A point in the WCS.

Returns

A 3D point.

Remarks

If the TextMovement property is set to **acDimLineWithText**, the dimension line will move along with the dimension text.

TextPrecision Property

Specifies the precision of angular dimensions.

```
(apa-get-TextPrecision Object)
(apa-set-TextPrecision Object Precision)
```

Arguments

Object	The DimAngular or Dim3PointAngular object to which this property applies.
Precision	An integer specifying the precision. One of the following:

`acDimPrecisionZero:`	`0`
`acDimPrecisionOne:`	`0.0`
`acDimPrecisionTwo:`	`0.00`
`acDimPrecisionThree:`	`0.000`
`acDimPrecisionFour:`	`0.0000`
`acDimPrecisionFive:`	`0.00000`
`acDimPrecisionSix:`	`0.000000`
`acDimPrecisionSeven:`	`0.0000000`
`acDimPrecisionEight:`	`0.00000000`

Returns

An integer.

Remarks

Overrides the DIMADEC system variable for the specified dimension.

TextPrefix Property

Specifies the dimension text prefix.

```
(apa-get-TextPrefix Object)
(apa-set-TextPrefix Object TextPrefix)
```

Arguments

Object	The Dim3PointAngular, DimAligned, DimAngular, DimDiametric, DimOrdinate DimRadial, or DimRotated object to which this property applies.
TextPrefix	A string, specifying the dimension text prefix.

Returns

An string.

Remarks

Overrides the DIMPOST system variable for the specified dimension.

An empty string (" ") specifies no prefix.

TextRotation Property

Specifies the rotation angle for dimension text.

```
(apa-get-TextRotation Object)
(apa-set-TextRotation Object TextRotation)
```

Arguments

Object	The Dim3PointAngular, DimAligned, DimAngular, DimDiametric, DimOrdinate DimRadial, or DimRotated object to which this property applies.
TextRotation	A number specifying the rotation angle in radians.

Returns

A real number in radians.

TextString Property

Specifies the text string for the object.

```
(apa-get-TextString Object)
(apa-set-TextString Object TextString)
```

Arguments

Object	The Attribute, AttributeRef, MText, Text, or Tolerance object to which this property applies.
TextString	A string.

Returns

A string.

TextStyle Property

Specifies the style for dimension text.

```
(apa-get-TextStyle Object)
(apa-set-TextStyle Object TextStyle)
```

Arguments

Object	The Dim3PointAngular, DimAligned, DimAngular, DimDiametric, DimOrdinate DimRadial, DimRotated, or Tolerance object to which this property applies.
TextStyle	A string, specifying the text style.

Returns

A real number in radians.

Remarks

Overrides the DIMTXSTY system variable for the specified dimension.

TextStyles Property

Returns the TextStyles collection for the document.

```
(apa-get-TextStyles Object)
```

Arguments

Object	The Document object to which this property applies.

Returns

The TextStyles collection for the document, as an object.

TextSuffix Property

Specifies the dimension text suffix.

```
(apa-get-TextSuffix Object)
(apa-set-TextSuffix Object TextSuffix)
```

Arguments

Object	The Dim3PointAngular, DimAligned, DimAngular, DimDiametric, DimOrdinate DimRadial, or DimRotated object to which this property applies.
TextSuffix	A string, specifying the dimension text suffix.

Returns

An string.

Remarks

Overrides the DIMPOST system variable for the specified dimension.

An empty string (" ") specifies no suffix.

TextureMapPath Property

Specifies the directories AutoCAD searches for texture maps.

```
(apa-get-TextureMapPath Object)
(apa-set-TextureMapPath Object Path)
```

Arguments

Object	The PreferencesFiles object to which this property applies.
Path	A string, specifying the directories in which to search.

Returns

A string.

Remarks

Semicolons separate multiple directories.

TextWinBackgrndColor Property

Specifies the background color of the AutoCAD text window.

```
(apa-get-TextWinBackgrndColor Object)
(apa-set-TextWinBackgrndColor Object Color)
```

Arguments

Object	The PreferencesDisplay object to which this method applies.
Color	An integer specifying an OLE_COLOR. One of the following:

> **vbBlack**
> **vbRed**
> **vbYellow**
> **vbGreen**
> **vbCyan**
> **vbBlue**
> **vbMagenta**
> **vbWhite**

Returns

An integer.

TextWinTextColor Property

Specifies the background color of the AutoCAD text window.

```
(apa-get-TextWinTextColor Object)
(apa-set-TextWinTextColor Object Color)
```

Arguments

Object	The PreferencesDisplay object to which this method applies.
Color	An integer specifying an OLE_COLOR. One of the following:

> **vbBlack**
> **vbRed**
> **vbYellow**
> **vbGreen**
> **vbCyan**
> **vbBlue**
> **vbMagenta**
> **vbWhite**

Returns

An integer.

Thickness Property

Specifies the extrusion thickness of 2D objects.

```
(apa-get-Thickness Object)
(apa-set-Thickness Object Thickness)
```

Arguments

Object | The Arc, Attribute, AttributeRef, Circle, LightweightPolyline, Line, Point, Polyline, Shape, Solid, Text, or Trace object to which this method applies.

Thickness | A number.

Returns

A real number.

ToleranceDisplay Property

Specifies the display of tolerances in dimensions.

```
(apa-get-ToleranceDisplay Object)
(apa-set-ToleranceDisplay Object DisplayMode)
```

Arguments

Object | The Dim3PointAngular, DimAligned, DimAngular, DimDiametric, DimOrdinate, DimRadial, or DimRotated object to which this method applies.

DisplayMode | A number. One of the following:

acTolNone
acTolSymmetrical
acTolDeviation
acTolLimits
acTolBasic

Returns

An integer.

Remarks

Overrides the DIMTOL system variable.

ToleranceHeightScale Property

Specifies the tolerance height scale factor for dimensions.

```
(apa-get-ToleranceHeightScale Object)
(apa-set-ToleranceHeightScale Object ScaleFactor)
```

Arguments

Object | The Dim3PointAngular, DimAligned, DimAngular, DimDiametric, DimOrdinate, DimRadial, or DimRotated object to which this method applies.

ScaleFactor | A non-zero, positive number.

Returns

An a real number.

Remarks

Overrides the DIMTFAC system variable.

ToleranceJustification Property

Specifies the vertical justification of tolerances in dimensions.

```
(apa-get-ToleranceJustification Object)
(apa-set-ToleranceJustification Object Justification)
```

Arguments

Object	The Dim3PointAngular, DimAligned, DimAngular, DimDiametric, DimOrdinate, DimRadial, or DimRotated object to which this method applies.
Justification	A number. One of the following:

`acTolTop`
`acTolMiddle`
`acTolBottom`

Returns

An integer.

Remarks

Overrides the DIMTOLJ system variable.

ToleranceLowerLimit Property

Specifies the negative tolerance for the dimension.

```
(apa-get-ToleranceLowerLimit Object)
(apa-set-ToleranceLowerLimit Object Limit)
```

Arguments

Object	The Dim3PointAngular, DimAligned, DimAngular, DimDiametric, DimOrdinate, DimRadial, or DimRotated object to which this method applies.
Limit	A number.

Returns

A real number.

Remarks

Overrides the DIMTM system variable.

TolerancePrecision Property

Specifies the precision of tolerance values in dimensions.

```
(apa-get-TolerancePrecision Object)
(apa-set-TolerancePrecision Object Precision)
```

Arguments

Object	The DimAligned, DimDiametric, DimOrdinate, DimRadial, or DimRotated object to which this property applies.
Precision	An integer specifying the precision. One of the following:

`acDimPrecisionZero:`	0
`acDimPrecisionOne:`	0.0
`acDimPrecisionTwo:`	0.00
`acDimPrecisionThree:`	0.000
`acDimPrecisionFour:`	0.0000
`acDimPrecisionFive:`	0.00000
`acDimPrecisionSix:`	0.000000
`acDimPrecisionSeven:`	0.0000000
`acDimPrecisionEight:`	0.00000000

Returns

An integer.

Remarks

Overrides the DIMTDEC system variable for the specified dimension.

ToleranceSuppressLeadingZeros Property

Specifies the suppression of leading zeros in tolerance values of dimensions.

```
(apa-get-ToleranceSuppressLeadingZeros Object)
(apa-set-ToleranceSuppressLeadingZeros Object Suppress)
```

Arguments

Object	The DimAligned, DimDiametric, DimOrdinate, DimRadial, or DimRotated object to which this property applies.
Suppress	non-nil to suppress leading zeros. nil to not-suppress.

Returns

T if leading zeros are suppressed. nil otherwise.

Remarks

Overrides the DIMTZIN system variable for the specified dimension.

ToleranceSuppressTrailingZeros Property

Specifies the suppression of trailing zeros in tolerance values of dimensions.

```
(apa-get-ToleranceSuppressTrailingZeros Object)
(apa-set-ToleranceSuppressTrailingZeros Object Suppress)
```

Arguments

Object	The DimAligned, DimDiametric, DimOrdinate, DimRadial, or DimRotated object to which this property applies.
Suppress	non-nil to suppress trailing zeros. nil to not-suppress.

Returns

T if trailing zeros are suppressed. nil otherwise.

Remarks

Overrides the DIMTZIN system variable for the specified dimension.

ToleranceSuppressZeroFeet Property

Specifies the suppression of a zero foot measurements in tolerance values of dimensions.

```
(apa-get-ToleranceSuppressZeroFeet Object)
(apa-set-ToleranceSuppressZeroFeet Object Suppress)
```

Arguments

Object	The DimAligned, DimDiametric, DimOrdinate, DimRadial, or DimRotated object to which this property applies.
Suppress	non-nil to suppress zero-feet. nil to not-suppress

Returns

T if zero feet are suppressed. nil otherwise.

Remarks

Overrides the DIMTZIN system variable for the specified dimension.

ToleranceSuppressZeroInches Property

Specifies the suppression of a zero inch measurement in tolerance values.

```
(apa-get-ToleranceSuppressZeroInches Object)
(apa-set-ToleranceSuppressZeroInches Object Suppress)
```

Arguments

Object	The DimAligned, DimDiametric, DimOrdinate, DimRadial, or DimRotated object to which this property applies.

Suppress   non-nil to suppress zero inches, nil to not-suppress.

Returns

T if zero inches are suppressed. nil otherwise.

Remarks

Overrides the DIMTZIN system variable for the specified dimension.

ToleranceUpperLimit Property

Specifies the positive tolerance for the dimension.

```
(apa-get-ToleranceUpperLimit Object)
(apa-set-ToleranceUpperLimit Object Limit)
```

Arguments

Object   The Dim3PointAngular, DimAligned, DimAngular, DimDiametric, DimOrdinate, DimRadial, or DimRotated object to which this method applies.

Limit   A number.

Returns

A real number.

Remarks

Overrides the DIMTP system variable.

Toolbars Property

Returns the toolbars collection for the specified menu group.

```
(apa-get-Toolbars Object)
```

Arguments

Object   The MenuGroup object to which this property applies.

Returns

An object.

Top Property

Specifies the top edge of a toolbar.

```
(apa-get-Top Object)
(apa-set-Top Object Top)
```

Arguments

Object   The Toolbar object to which this property applies.

Top   A number specifying the top edge of the toolbar in pixels.

Returns

An integer.

TotalAngle Property

Returns the enclosed angle an arc.

```
(apa-get-StartAngle Object)
```

Arguments

Object   The Arc object to which this method applies.

Returns

A real number, in radians.

Transparency Property

Specifies the transparency of a bi-tonal raster image.

```
(apa-get-Transparency Object)
(apa-set-Transparency Object Transparency)
```

Arguments

Object   The bi-tonal Raster object to which this property applies.

Transparency   non-nil specifies the object is transparent, nil that it isn't.

Returns

T if the object is transparent, nil if it isn't

TrueColorImages Property

Controls the display of TrueColor rendering and images.

```
(apa-get-TrueColorImages Object)
(apa-set- TrueColorImages Object Display)
```

Arguments

Object   The PreferencesDisplay object to which this property applies.

Display   non-nil to display in TrueColor, nil to not.

Returns

T if displayed in TrueColor, nil if not.

TwistAngle Property

Specifies the twist angle of the viewport.

```
(apa-get-TwistAngle Object)
(apa-set-TwistAngle Object Angle)
```

Arguments

Object   The PViewport object to which this property applies.

Angle   A number specifying the twist angle for the viewport in radians.

Returns

A real number.

Type Property

Specifies the type of the specified object.

```
(apa-get-Type Object)
(apa-set-Type Object Type)
```

Arguments

Object   The 3DPoly, Layout, Leader, MenuGroup, PlotConfiguration, Polyline, PolygonMesh, PopupMenuItem, or ToolbarItem object to which this property applies.

Type   A number. For a Leader, one of the following:

```
acLineNoArrow
acLineWithArrow
acSplineNoArrow
acSplineWithArrow
```

For a MenuGroup, one of the following:

acBaseMenuGroup
acPartialMenuGroup

For a PopupMenuItem, one of the following:

acMenuItem

```
acMenuSeparator
acMenuSubMenu.
```
For a ToolbarItem, one of the following:
```
acToolbarButton
acToolbarFlyout
acToolbarControl
acToolbarSeparator
```
For a 3DPoly, one of the following:
```
acSimple3DPoly.
acQuadSpline3DPoly
acCubicSpline3DPoly.
```
For a Polyline, one of the following:
```
acSimplePoly
acFitCurvePoly.
acQuadSplinePoly
acCubicSplinePoly
```
For a PolygonMesh, one of the following:
```
acSimpleMesh
acQuadSurfaceMesh
acCubicSurfaceMesh
acBezierSurfaceMesh
```
For a Layout or a PlotConfiguration, one of the following:
```
acDisplay
acExtents
acLimits
acView
acWindow
acLayout
```

Returns

An integer.

UCSIconAtOrigin Property

Controls the display of the UCS icon.
```
(apa-get-UCSIconAtOrigin Object)
(apa-set-UCSIconAtOrigin Object AtOrigin)
```

Arguments

Object	The PViewport or Viewport object to which this property applies.
AtOrigin	non-nil to display the UCS icon at the origin of the UCS, nil to not display it there.

Returns

T if the UCS icon is displayed at the origin, nil if it's not.

Remarks

You must reset the active viewport to see the changes.

UCSIconOn Property

Controls the display of the UCS icon.
```
(apa-get-UCSIconOn Object)
(apa-set-UCSIconOn Object On)
```

Arguments

Object	The PViewport or Viewport object to which this property applies.
On	non-nil to display the UCS icon, nil to not display it.

Returns

T if the UCS icon is displayed, nil if it's not.

You must reset the active viewport to see the changes.

UCSPerViewport Property

Specifies if the UCS is saved with the viewport.

```
(apa-get-UCSPerViewport Object)
(apa-set-UCSPerViewport Object UCSPerViewport)
```

Arguments

Object   The PViewport object to which this property applies.

UCSPerViewport   non-nil to save the UCS with the viewport, nil to not.

Returns

T if the UCS is saved with the viewport, nil if it's not.

UnitsFormat Property

Specifies the linear units format for dimensions.

```
(apa-get-UnitsFormat Object)
(apa-set-UnitsFormat Object UnitsFormat)
```

Arguments

Object   The DimAligned, DimDiametric, DimOrdinate, DimRadial, or DimRotated object to
                    which this property applies.

Format   An integer specifying the format. One of the following:

                    acDimScientific
                    acDimDecimal
                    acDimEngineering
                    acDimArchitecturalStacked
                    acDimFractionalStacked
                    acDimArchitectural
                    acDimFractional
                    acDimWindowsDesktop

Returns

An integer.

Remarks

Overrides the DIMLUNIT system variable for the specified dimension.

UpperRightCorner Property

Returns the upper-right corner of the specified viewport.

```
(apa-get-UpperRightCorner Object)
```

Arguments

Object   The Viewport object to which this property applies.

Returns

A 2D point representing the upper-right corner of the viewport.

UpsideDown Property

Specifies the direction of text.

```
(apa-get-UpsideDown Object)
(apa-set-UpsideDown Object Mode)
```

Arguments

Object	The Attribute, AttributeReference, or Text object to which this property applies.
Mode	non-nil for upside-down, nil for upside-up.

Returns

T if upside-down. nil if upside-up.

URL Property

Specifies the URL for a Hyperlink object.

```
(apa-get-URL Object)
(apa-set-URL Object URL)
```

Arguments

Object	The Hyperlink object to which this property applies.
URL	A string.

Returns

A string.

URLDescription Property

Specifies the URLDescription for a Hyperlink object.

```
(apa-get-URLDescription Object)
(apa-set-URLDescription Object URLDescription)
```

Arguments

Object	The Hyperlink object to which this property applies.
URLDescription	A string.

Returns

A string.

URLNamedLocation Property

Specifies the Named Location for a Hyperlink object.

```
(apa-get-URLNamedLocation Object)
(apa-set-URLNamedLocation Object URLNamedLocation)
```

Arguments

Object	The Hyperlink object to which this property applies.
URLNamedLocation	A string.

Returns

A string.

UseLastPlotSettings Property

Specifies use of the last successful plot settings.

```
(apa-get-UseLastPlotSettings Object)
(apa-set-UseLastPlotSettings Object Use)
```

Arguments

Object	The PreferencesOutput object to which this property applies.
Use	non-nil to use the last successful plot settings. nil to not.

Returns

T if using the last successful plot settings, nil if not.

User Property

Returns the PreferencesUser collection for the document.

`(apa-get-Users Object)`

Arguments

`Object`   The Preferences object to which this property applies.

Returns

The PreferencesUser collection for the document, as an object.

Remarks

There are two types of options on the User Preferences tab of the Options dialog box:

- Those that are stored in the current drawing, affecting only that drawing. These are held in the DatabasePreferences object.

- Those that are stored in the system registry, affecting the AutoCAD application. These are held in the PreferencesUser object.

The following options from the User Preferences tab are held in the DatabasePreferences object: `ContourLinesPerSurface`, `DisplaySilhouette`, `Lineweight`, `LineWeightDisplay`, `ObjectSortByPlotting`, `ObjectSortByPSOutput`, `ObjectSortByRedraws`, `ObjectSortByRegens`, `ObjectSortBySelection`, and `ObjectSortBySnap`.

UserCoordinateSystems Property

Returns the UserCoordinateSystems collection for the document.

`(apa-get-UserCoordinateSystems Object)`

Arguments

`Object`   The Document object to which this property applies.

Returns

The UserCoordinateSystems collection for the document, as an object.

UseStandardScale Property

Specifies the use of a standard scale with a plot.

`(apa-get-UseStandardScale Object)`
`(apa-set-UseStandardScale Object StandardScale)`

Arguments

`Object`   The Layout or PlotConfiguration object to which this property applies.
`StandardScale`   `non-nil` to a standard scale. `nil` to use a custom scale.

Returns

`T` if using a standard scale, `nil` if using a custom scale.

Utility Property

Returns the utility object for the document.

`(apa-get-Utility Object)`

Arguments

`Object`   The Document object to which this property applies.

Returns

The utility object.

Verify Property

Specifies if the attribute or attribute reference is verified.

```
(apa-get-Verify Object)
(apa-set-Verify Object Verify)
```

Arguments

Object   The Attribute or AttributeReference object to which this property applies.

Verify   non-nil specifies the object is verified, nil that it isn't.

Returns

T if the object is verified, nil if it isn't.

Version Property

Returns the version of AutoCAD that you're using.

```
(apa-get-Version Object)
```

Arguments

Object   The Application object to which this property applies.

Returns

A string.

Remarks.

Stored in the ACADVER system variable.

The initial release of AutoCAD 2000 returns "15.0".

VerticalTextPosition Property

Specifies the vertical alignment for dimension text.

```
(apa-get-VerticalTextPosition Object)
(apa-set-VerticalTextPosition Object Alignment)
```

Arguments

Object   The Dim3PointAngular, DimAligned, DimAngular, DimDiametric, DimRadial, or
                     DimRotated, or Leader object to which this property applies.

Alignment   An integer specifying the alignment. One of the following:

acVertCentered
acAbove
acOutside
acJIS

Returns

The alignment as an integer.

Remarks

Overrides the DIMTAD system variable for the given dimension.

ViewportDefault Property

Specifies if the layer is to be frozen in new viewports.

```
(apa-get-ViewportDefault Object)
(apa-set-ViewportDefault Object Freeze)
```

Arguments

Object   The Layer object to which this property applies.

Freeze   non-nil to freeze in new viewports; nil to not freeze.

Returns

T if frozen in new viewports, nil if not.

ViewportOn Property

Specifies the status of the viewport.

```
(apa-get-ViewportOn Object)
(apa-set-ViewportOn Object On)
```

Arguments

Object   The PViewport object to which this property applies.

On   non-nil to turn it on; nil to turn it off.

Returns

T if on, nil if not.

Viewports Property

Returns the Viewports collection for the document.

```
(apa-get-Viewports Object)
```

Arguments

Object   The Document object to which this property applies.

Returns

The Viewports collection for the document, as an object.

Views Property

Returns the Views collection for the document.

```
(apa-get-Views Object)
```

Arguments

Object   The Document object to which this property applies.

Returns

The Views collection for the document, as an object.

ViewToPlot Property

Specifies the view to plot for a layout or plot configuration.

```
(apa-get-ViewToPlot Object)
(apa-set-ViewToPlot Object ViewToPlot)
```

Arguments

Object   The Layout or PlotConfiguration object to which this property applies.

String   The name of the named view to plot.

Returns

A string.

Visible Property

Specifies the visibility of the object.

```
(apa-get-Visible Object)
(apa-set-Visible Object Visible)
```

Arguments

Object   The All Drawing objects, Application, AttributeRef, Group, or Toolbar object to which this property applies.

Visible　　　　　non-nil to turn it on; nil to turn it off.

Returns

T if visible, nil if not.

VisibilityEdge1 Property

Specifies the visibility of edge 1 of a 3D face object.

```
(apa-get-VisibilityEdge1 Object)
(apa-set-VisibilityEdge1 Object Visible)
```

Arguments

Object　　　　　The 3DFace object to which this property applies.

Visible　　　　　non-nil to turn it on; nil to turn it off.

Returns

T if visible, nil if not.

VisibilityEdge2 Property

Specifies the visibility of edge 2 of a 3D face object.

```
(apa-get-VisibilityEdge2 Object)
(apa-set-VisibilityEdge2 Object Visible)
```

Arguments

Object　　　　　The 3DFace object to which this property applies.

Visible　　　　　non-nil to turn it on; nil to turn it off.

Returns

T if visible, nil if not.

VisibilityEdge3 Property

Specifies the visibility of edge 3 of a 3D face object.

```
(apa-get-VisibilityEdge3 Object)
(apa-set-VisibilityEdge3 Object Visible)
```

Arguments

Object　　　　　The 3DFace object to which this property applies.

Visible　　　　　non-nil to turn it on; nil to turn it off.

Returns

T if visible, nil if not.

VisibilityEdge4 Property

Specifies the visibility of edge 4 of a 3D face object.

```
(apa-get-VisibilityEdge1 Object)
(apa-set-VisibilityEdge1 Object Visible)
```

Arguments

Object　　　　　The 3DFace object to which this property applies.

Visible　　　　　non-nil to turn it on; nil to turn it off.

Returns

T if visible, nil if not.

Volume Property

Returns the volume of a 3DSolid.

`(apa-get-Volume Object)`

Arguments

Object   The 3DSolid object to which this property applies.

Returns

A real number.

Weights Property

Specifies the weight vector for a spline.

`(apa-get-Weights Object)`
`(apa-set-Weights Object Weights)`

Arguments

Object   The Spline object to which this property applies.

Returns

The weight vector of the spline.

Remarks

You must use the SetWeight method at least once before getting or setting the weights property.

Width Property

Specifies the width of the main application window, attribute, raster, shape, text, toolbar, view, or viewport.

`(apa-get-Width Object)`
`(apa set-Width Object Width)`

Arguments

Object   The Application, Attribute, AttributeRef, MText, PViewport, Raster, Shape, Text,
                       TextStyle, Toolbar, Viewport, or View object to which this property applies.

Width   A positive number.
                       In pixels, for Application, Raster and Toolbar objects.
                       The horizontal width factor for TextStyles and Text.
                       In current drawing units for all others.

Returns

An integer, in pixels, in pixels, for Application, Raster, and Toolbar objects.

A real number for all others.

WindowLeft Property

Specifies the left edge of the application window in pixels.

`(apa-get-WindowLeft Object)`
`(apa-set-WindowLeft Object Left)`

Arguments

Object   The Application object to which this property applies.

Left   A positive number specifying the left edge of the application window in pixels.

Returns

An integer.

WindowState Property

Specifies the state of the application or document window.

```
(apa-get-WindowState Object)
(apa-set-WindowState Object WindowState)
```

Arguments

Object	The Application or Document object to which this property applies.
Left	One of the following:

acMin	Minimized.
acMax	Maximized.
acNorm	Normal

Returns

An integer.

WindowTitle Property

Returns the title of the document window.

```
(apa-get-WindowTitle Object)
```

Arguments

Object	The Document object to which this property applies.

Returns

A string.

WindowTop Property

Specifies the top edge of the application window in pixels.

```
(apa-get-WindowTop Object)
(apa-set-WindowTop Object Top)
```

Arguments

Object	The Application object to which this property applies.
Top	A positive number specifying the top edge of the application window in pixels.

Returns

An integer.

WorkspacePath Property

Specifies the directory AutoCAD uses for the database workspace file.

```
(apa-get-WorkspacePath Object)
(apa-set-WorkspacePath Object Path)
```

Arguments

Object	The PreferencesFiles object to which this property applies.
Path	A string, specifying the drive and path of the database workspace file.

Returns

A string.

XRefDatabase Property

Returns the database object that defines a block.

```
(apa-get-XRefDatabase Object)
```

Arguments

Object	The Block object to which this property applies.

Returns

The database object that defines the block.

XRefDemandLoad Property

Controls the demand loading of XRefs.

```
(apa-get-XRefDemandLoad Object)
(apa-set-XRefDemandLoad Object DemandLoad)
```

Arguments

Object	The PreferencesOpenSave object to which this property applies.
DemandLoad	An integer. One of the following: **acDemandLoadDisabled acDemandLoadEnabled acDemandLoadEnabledWithCopy**

Returns

An integer.

Remarks

This property is stored in the XLOADCTL system variable.

XRefEdit Property

Specifies if the specified drawing, when XRef'd, can be edited in place.

```
(apa-get-XRefEdit Object)
(apa-set-XRefEdit Object Enable)
```

Arguments

Object	The DatabasePreferences object to which this property applies.
Enable	non-nil to enable editing in place, nil to disable.

Returns

T if enabled, nil if disabled.

Remarks

This property is stored in the XEDIT system variable.

XRefFadeIntensity Property

Controls the fading intensity for XRef's being edited in-place.

```
(apa-get-XRefFadeIntensity Object)
(apa-set-XRefFadeIntensity Object Fade)
```

Arguments

Object	The PreferencesDisplay object to which this property applies.
Fade	An integer in the range $0 \leq Fade \leq 90$, specifying the fade of the image.

Returns

An integer.

Remarks

This property is stored in the XFADECTL system variable. The default value is 50.

XRefLayerVisibility Property

Specifies if changes to xref-dependent layers and paths are saved in the current drawing.

```
(apa-get-XRefLayerVisibility Object)
(apa-set-XRefLayerVisibility Object Saved)
```

Arguments

Object	The DatabasePreferences object to which this property applies.
Saved	non-nil to save changes, nil to discard.

Returns

T if saved, nil if discarded.

Remarks

This property is stored in the VISRETAIN system variable.

XScaleFactor Property

Specifies the X Scale factor for the object.

```
(apa-get-XScaleFactor Object)
(apa-set-XScaleFactor Object Factor)
```

Arguments

Object	The BlockRef, ExternalReference, or MInsertBlock object to which this property applies.
Factor	A non-zero number.

Returns

A real number.

XVector Property

Specifies the X direction of the UCS

```
(apa-get-XVector Object)
(apa-set-XVector Object Vector)
```

Arguments

Object	The UCS object to which this property applies.
Vector	A 2D or 3D vector.

Returns

A 3D unit vector.

Remarks

This property is stored in the UCSXDIR system variable.

Changes to this property for the active UCS will not be visible until you reset the active UCS.

YScaleFactor Property

Specifies the Y Scale factor for the object.

```
(apa-get-YScaleFactor Object)
(apa-set-YScaleFactor Object Factor)
```

Arguments

Object	The BlockRef, ExternalReference, or MInsertBlock object to which this property applies.
Factor	A non-zero number.

Returns

A real number.

A

M

N

LICENSE AGREEMENT FOR AUTODESK PRESS

THOMSON LEARNING™

Educational Software/Data

You the customer, and Autodesk Press incur certain benefits, rights, and obligations to each other when you open this package and use the software/data it contains. BE SURE YOU READ THE LICENSE AGREEMENT CAREFULLY, SINCE BY USING THE SOFTWARE/DATA YOU INDICATE YOU HAVE READ, UNDERSTOOD, AND ACCEPTED THE TERMS OF THIS AGREEMENT.

Your rights:

1. You enjoy a non-exclusive license to use the enclosed software/data on a single microcomputer that is not part of a network or multi-machine system in consideration for payment of the required license fee, (which may be included in the purchase price of an accompanying print component), or receipt of this software/data, and your acceptance of the terms and conditions of this agreement.

2. You own the media on which the software/data is recorded, but you acknowledge that you do not own the software/data recorded on them. You also acknowledge that the software/data is furnished "as is," and contains copyrighted and/or proprietary and confidential information of Autodesk Press or its licensors.

3. If you do not accept the terms of this license agreement you may return the media within 30 days. However, you may not use the software during this period.

There are limitations on your rights:

1. You may not copy or print the software/data for any reason whatsoever, except to install it on a hard drive on a single microcomputer and to make one archival copy, unless copying or printing is expressly permitted in writing or statements recorded on the diskette(s).

2. You may not revise, translate, convert, disassemble or otherwise reverse engineer the software/data except that you may add to or rearrange any data recorded on the media as part of the normal use of the software/data.

3. You may not sell, license, lease, rent, loan, or otherwise distribute or network the software/data except that you may give the software/data to a student or and instructor for use at school or, temporarily at home.

Should you fail to abide by the Copyright Law of the United States as it applies to this software/data your license to use it will become invalid. You agree to erase or otherwise destroy the software/data immediately after receiving note of Autodesk Press' termination of this agreement for violation of its provisions.

Autodesk Press gives you a LIMITED WARRANTY covering the enclosed software/data. The LIMITED WARRANTY can be found in this product and/or the instructor's manual that accompanies it.

This license is the entire agreement between you and Autodesk Press interpreted and enforced under New York law.

Limited Warranty

Autodesk Press warrants to the original licensee/ purchaser of this copy of microcomputer software/ data and the media on which it is recorded that the media will be free from defects in material and workmanship for ninety (90) days from the date of original purchase. All implied warranties are limited in duration to this ninety (90) day period. THEREAFTER, ANY IMPLIED WARRANTIES, INCLUDING IMPLIED WARRANTIES OF MERCHANTABILITY AND FITNESS FOR A PARTICULAR PURPOSE ARE EXCLUDED. THIS WARRANTY IS IN LIEU OF ALL OTHER WARRANTIES, WHETHER ORAL OR WRITTEN, EXPRESSED OR IMPLIED.

If you believe the media is defective, please return it during the ninety day period to the address shown below. A defective diskette will be replaced without charge provided that it has not been subjected to misuse or damage.

This warranty does not extend to the software or information recorded on the media. The software and information are provided "AS IS." Any statements made about the utility of the software or information are not to be considered as express or implied warranties. Autodesk Press will not be liable for incidental or consequential damages of any kind incurred by you, the consumer, or any other user.

Some states do not allow the exclusion or limitation of incidental or consequential damages, or limitations on the duration of implied warranties, so the above limitation or exclusion may not apply to you. This warranty gives you specific legal rights, and you may also have other rights which vary from state to state. Address all correspondence to:

Autodesk Press
3 Columbia Circle
P. O. Box 15015
Albany, NY 12212-5015